P9-AQY-369

MICHAEL NACHT

The Age of Vulnerability

Threats to the Nuclear Stalemate

THE BROOKINGS INSTITUTION
Washington, D.C.

1775 Massachusetts Avenue, N.W., Washington, D.C. 20036

Library of Congress Cataloging in Publication data:

Nacht, Michael.
The age of vulnerability.
Includes bibliographical references and index.
1. Nuclear weapons—United States. 2. Nuclear weapons —Soviet Union. 3. United States—Military relations— Soviet Union. 4. Soviet Union—Military relations— United States. 5. World Politics—1975–1985. I. Title.
U264.N3 1985 355′.0330047 84-45849
ISBN 0-8157-5964-9
ISBN 0-8157-5963-0 (pbk.)

1 2 3 4 5 6 7 8 9

THE BROOKINGS INSTITUTION is an independent organization devoted to nonpartisan research, education, and publication in economics, government, foreign policy, and the social sciences generally. Its principal purposes are to aid in the development of sound public policies and to promote public understanding of issues of national importance.

The Institution was founded on December 8, 1927, to merge the activities of the Institute for Government Research, founded in 1916, the Institute of Economics, founded in 1922, and the Robert Brookings Graduate School of Economics and Government, founded in 1924.

The Board of Trustees is responsible for the general administration of the Institution, while the immediate direction of the policies, program, and staff is vested in the President, assisted by an advisory committee of the officers and staff. The by-laws of the Institution state: "It is the function of the Trustees to make possible the conduct of scientific research, and publication, under the most favorable conditions, and to safeguard the independence of the research staff in the pursuit of their studies and in the publication of the results of such studies. It is not a part of their function to determine, control, or influence the conduct of particular investigations or the conclusions reached."

The President bears final responsibility for the decision to publish a manuscript as a Brookings book. In reaching his judgment on the competence, accuracy, and objectivity of each study, the President is advised by the director of the appropriate research program and weighs the views of a panel of expert outside readers who report to him in confidence on the quality of the work. Publication of a work signifies that it is deemed a competent treatment worthy of public consideration but does not imply endorsement of conclusions or recommendations.

The Institution maintains its position of neutrality on issues of public policy in order to safeguard the intellectual freedom of the staff. Hence interpretations or conclusions in Brookings publications should be understood to be solely those of the authors and should not be attributed to the Institution, to its trustees, officers, or other staff members, or to the organizations that support its research.

To my mother and father,
Ann and Jack Nacht,
for their love and inspiration

Foreword

DURING the last decade, as U.S.-Soviet strategic competition has intensified with the collapse of détente, the fear of nuclear war between the superpowers has become a prominent national and international concern. The ensuing public debate has proceeded along several paths: the assessment of Soviet intentions, the future of ballistic missile defense, the evolution of NATO's nuclear strategy and the feasibility of strengthening its conventionl forces, and the prospects for nuclear arms control, among others.

In this book, Michael Nacht takes an intentionally broad perspective. He seeks to clarify and integrate the principal roles played by nuclear weapons in the national security policies of both the United States and the Soviet Union. In his analysis he is concerned with the structural constraints on the superpowers' nuclear competition and the circumstances under which these constraints would be lifted. He therefore concentrates on those aspects of Soviet and American political culture, military technology and doctrine, alliance relationships, and arms control that relate directly to this long-term perspective.

Nacht, who now teaches national security policy and public management at the University of Maryland's School of Public Affairs, completed much of the work on this study while he was teaching at Harvard University's Center for Science and International Affairs. He is grateful to the director of the center, Paul Doty, and to George Eads and Catherine Kelleher of the University of Maryland for their support. A Senior Scholar Traveling Fellowship to the Soviet Union, provided by the International Exchange Board in 1983, helped to shape the author's understanding of Soviet perspectives.

The author wishes to thank the many people who offered constructive comments on his manuscript: Barry M. Blechman, Harvey Brooks,

Crauford D. Goodwin, William Hyland, Richard E. Neustadt, Joseph Nye, Jack Ruina, Thomas Schelling, John D. Steinbruner, and, in particular, Ted Greenwood, Steven Miller, and David Nacht. He also thanks his wife, Marjorie Jo, for her unflagging encouragement and support. Jean Rosenblatt edited the manuscript; Nancy A. Ameen and Alan G. Hoden verified its factual content; Diane Asay, Michelle Marcouiller, Mary Ann Wells, and especially Lynn Page Whittaker provided secretarial and editorial support; and Ward & Sylvan prepared the index.

The study was funded in part by a grant from the Ford Foundation. Parts of chapters 3, 5, and 8 were adapted by the author from articles published in *Daedalus* (Winter 1980–81); *Foreign Policy* (Spring 1982); *Naval War College Review* (November–December 1983); *International Organization* (Winter 1980–81); and *Annals of the American Academy of Political and Social Science* (March 1977). Parts of chapters 6 and 9 were first developed in a paper commissioned by the Humphrey Institute of the University of Minnesota.

The views in this book are those of the author and should not be ascribed to the persons or organizations whose assistance is acknowledged above, or to the trustees, officers, or staff members of the Brookings Institution.

BRUCE K. MAC LAURY
President

February 1985
Washington, D.C.

Contents

CHAPTER ONE

Introduction

The hallmark of responsible comment is not to sit in judgment on events as an idle spectator. . . . Responsibility consists in sharing the burden on men directing what is to be done or the burden of offering some other course of action in the mood of one who has realized what it would mean to undertake it.

Walter Lippmann

WE LIVE in an age of vulnerability, a historical period marked by an intense and dangerous competition between two military giants, the United States and the Soviet Union, each possessing significant political, economic, and military weaknesses that the other is seeking to exploit. The United States, striving to regain its national self-confidence in the aftermath of the Vietnam War, is attempting to contain Soviet expansionism at a time when the overall quantitative military balance of forces remains highly unfavorable to the West, when the political unity of the North Atlantic Treaty Organization (NATO) continues to show signs of serious strain, and when massive budget deficits plague the American economic system. The Soviet Union, on the other hand, is seeking to derive concrete political and economic benefits from the massive investment in military forces it made in the 1960s and 1970s, while simultaneously trying to prevent the formation of a U.S.-European-Chinese-Japanese alliance, which would effectively encircle the Soviet Union. However, the Soviet leadership is seriously hampered in achieving its foreign policy objectives by the burden of the empire it has amassed, especially in Eastern Europe, and by the chronic inefficiencies of an economic system that, by the early 1980s, was yielding extremely modest growth rates.

Central to what determines the competition between the two powers is their nuclear forces. After the United States and the Soviet Union emerged from the Second World War as the dominant countries in international politics, nuclear weapons and then their delivery vehicles became the most visible symbols of superpower status. Ever since, they have been the principal political litmus test for judging where advantage resides in Soviet-American relations. Indeed, the widespread belief among Americans by the end of the 1970s that the military advantage had shifted to the Soviet Union, despite negotiated nuclear arms control agreements, was what destroyed the domestic political support for an American policy that emphasized cooperation—détente—with the Soviet Union.

When the United States and the Soviet Union first reached agreement at the strategic arms limitation talks (SALT) in May 1972, the view was widely held that the stability of the nuclear balance between the superpowers had been enhanced significantly and that an important step had been taken toward maintaining that stability. Most American observers seemed to believe that the treaty to limit the deployment of ballistic missile defenses to two sites (later reduced to one) virtually ensured that both countries would lack confidence in their ability to survive a retaliatory attack, thereby precluding serious consideration of a first strike, even in a crisis. An interim agreement on offensive weapons was more problematic, though, because it codified the Soviet advantage in the number of strategic missile launchers it had already deployed. But observers expected that agreement to be soon replaced by a more durable one that would effectively constrain all the long-range nuclear force deployments of both powers. It was hoped that these agreements (SALT I) and anticipated follow-up measures would transform Soviet-American relations, making them cooperative rather than competitive.

In testifying before Congress on the merits of the SALT I agreements, Henry A. Kissinger, then the assistant to the president for national security affairs, offered the following observation:

> The final verdict must wait on events, but there is at least reason to hope that these accords represent a major break in the pattern of suspicion, hostility, and confrontation which has dominated U.S.-Soviet relations for a generation.[1]

By the late 1970s the guarded optimism of 1972 had been replaced by a deep skepticism of the feasibility of building such a cooperative relationship. The usefulness of the SALT I agreements and the desirability of continuing arms control negotiations were seriously questioned. President Jimmy Carter withdrew the SALT II Treaty—criticized by some as perpetuating a senseless arms race and by others as codifying Soviet strategic superiority—from the Senate ratification process following the Soviet invasion of Afghanistan in December 1979. Subse-

1. *Strategic Arms Limitation Agreements,* Hearings before the Senate Committee on Foreign Relations, 92 Cong. 2 sess. (Government Printing Office, 1972), p. 400.

quently President Ronald Reagan called the treaty "fatally flawed," and it was abandoned in favor of a negotiating approach that ostensibly emphasized strategic arms reductions, although both the United States and the Soviet Union declared at the time that they would adhere to the terms of SALT II.

Between the time the SALT I agreements entered into force in October 1972 and President Reagan's election in November 1980, much had happened to persuade Americans that the Soviet Union posed a far greater threat to the security of the United States and to the West than the détente policies of Presidents Nixon, Ford, and Carter had led them to believe.

Several factors brought about this change in attitude, which collectively represented the "conservative critique" of American national security policy espoused by President Reagan. First, the vigorous modernization program of Soviet intercontinental ballistic missile (ICBM) systems produced a force that was a serious threat to the survivability of U.S. land-based missiles, a situation, it was argued, that the Soviet leadership could perhaps exploit to achieve political or even military gain.

Second, numerical indicators comparing Soviet and American nuclear forces favored the Soviet Union in most cases. When the SALT I agreements entered into force, the United States enjoyed a noteworthy advantage in the number and accuracy of deployed warheads. By the time the SALT II Treaty was signed, the Soviet Union had surpassed the United States in numbers of warheads and many other measures.[2]

Third, official U.S. estimates of Soviet defense expenditures as a percentage of its gross national product were revised upward several times in the 1970s, suggesting that the United States had substantially underestimated the scope and magnitude of Soviet resources allocated to military programs.

Fourth, Soviet violations of the SALT I accords were alleged to have occurred on several occasions, raising serious doubts about Soviet intentions and the wisdom of entering into bilateral agreements with the

2. A study conducted by the Department of Defense in 1984 concluded that as of that time the Soviet Union had deployed about 34,000 nuclear warheads for its bombers, long-range and medium-range missiles, and artillery and cruise missiles, whereas, by comparison, the United States had deployed 26,000 warheads. The study claimed that the Soviet Union overtook the United States in nuclear warheads in 1978–79. See Richard Halloran, "Soviet Said to Lead U.S. by 8,000 Warheads," *New York Times*, June 18, 1984.

Soviet government. Moreover, intelligence information gathered from incidents at Sverdlovsk in the Soviet Union and from Laos, Cambodia, and Afghanistan cast doubt on Soviet compliance with the Biological Weapons Convention in the former instance and with the Geneva Protocol prohibiting wartime use of poisonous gases in the latter.

Fifth, evidence of an extensive Soviet civil defense program strengthened the contention that Soviet leaders rejected deterrence as defined by American strategists and that these leaders were planning to wage nuclear war and prevail. Critics of SALT II claimed that a "war survivability" gap had developed between the Soviet Union and the United States and that the former would be in a far better position than the latter to recover from the effects of a nuclear war and to dominate the peace that would follow.

Sixth, the Soviet Union was found to be conducting vigorous research, development, and testing programs in the application of exotic technologies (for example, directed-energy systems such as high-energy lasers and particle beams) for both ballistic missile defense (BMD) systems and for antisatellite (ASAT) weapon systems.

Seventh, Soviet military writings emphasized doctrines for fighting nuclear war and did not endorse concepts of strategic balance, deterrence, or stability as defined by American writers and policymakers. Some American analysts deduced malevolent Soviet intentions from these writings, including Soviet plans to fight a nuclear war.

Finally, Soviet deployments starting in the mid-1970s of a new intermediate-range ballistic missile (IRBM), the SS-20, and a highly advanced fighter-bomber known in the West as the Backfire supported the view that nuclear forces in Europe, intended to defend member states of NATO, were increasingly vulnerable to attack. Concern developed that a Soviet preemptive strike using advanced systems could prevent NATO from using nuclear weapons in response to an initial attack by conventional forces of the Warsaw Pact countries.

Through the 1970s Americans became more distrustful of the objectives of Soviet foreign policy and more concerned that the Soviet Union would use its military power to achieve these objectives.[3] This shift in attitudes was based not only on the nuclear arsenals of the two powers but also on conventional forces. In the 1970s the Soviet Union worked

3. For supporting evidence see John E. Reilly, ed., *American Public Opinion and U.S. Foreign Policy, 1979* (Chicago Council on Foreign Relations, 1979), particularly pp. 25–27.

to become a global military power. Nonnuclear Soviet forces allocated to the European theater increased modestly in numbers and improved markedly in capabilities, causing the West to become anxious about the prospects of a Warsaw Pact attack on NATO territory. The transition of the Soviet navy from a force capable only of coastal defense to a true "blue water" navy provided the Soviet Union, for the first time, with the ability to threaten U.S. sea lines of communication and to project political power far from the Soviet homeland. Finally, the Soviet Union's growing airlift-sealift capability, demonstrated most notably in Ethiopia in 1977, meant that Soviet military intervention in a wide range of regional crises was fast becoming a feasible option.

The Soviet Union combined its significant growth in military power with an aggressive foreign policy that took many in the West by surprise. For example, the Soviet Union helped North Vietnam violate the 1973 Paris Peace Accords, contributing eventually to the collapse of South Vietnam and to the establishment of a unified communist Vietnamese state. Soviet behavior in the October 1973 war in the Middle East was contrary to the American understanding of the meaning of superpower détente. The Soviets promoted introduction of Cuban troops and military advisers into the Angolan civil war and subsequently into several other countries in sub-Saharan Africa. They encouraged communist coups d'état in Afghanistan and South Yemen. In December 1979 Soviet forces invaded and occupied Afghanistan, the first combat role for Soviet troops outside Eastern Europe since World War II.

Moreover, the Soviet Union's commitment to respect human rights embodied in its acceptance of the 1975 agreement at the Conference on Security and Cooperation in Europe (the so-called Helsinki Agreement) was in stark contrast to Soviet government behavior. The writings of Alexander Solzhenitsyn and Andrei Sakharov, the celebrated trials of Alexander Ginsberg and other critics of the regime, and the arrest and harassment in Moscow of American businessmen and journalists all portrayed an image of Soviet society that was inimical to Western values and traditions.

But the tendency toward greater conservatism, reflected in widespread support for large increases in U.S. defense spending, was not shared by all Americans in the early 1980s. Many offered to counter the points on which the case for conservatism rested. It was emphasized that the problem of U.S. land-based missile vulnerability was theoretical at best and that the Soviet Union could not exploit that vulnerability for material gain. Some analysts pointed out that in the overall balance of nuclear forces the superpowers had achieved rough parity, and both had far more weapons than were required to achieve any sensible military objective. Indeed, during President Reagan's first term a "nuclear freeze" movement developed that commanded widespread public support. That the Soviets spent much more on defense than was previously thought, skeptics observed, told us more about the inefficiencies of the Soviet economy than it did about actual Soviet capabilities or intentions. Some defense specialists claimed that there was no incontrovertible evidence supporting the allegations of Soviet violations of arms control agreements, that the Soviet civil defense program reflected more the traditional Russian penchant for defense and population control than it revealed anything about Soviet plans to fight a nuclear war or ability to survive it, and that the significance of Soviet research on directed-energy systems was highly exaggerated.

Some Sovietologists asserted that Soviet military writings stressing war fighting were the natural product of men whose job it was to prepare to fight and that these views did not at all represent the attitudes of Soviet political leaders, who had a healthy respect for nuclear deterrence. Experts claimed that the Soviet deployment of new nuclear systems for use in Europe did not really affect the likelihood of a war there, because U.S. sea-based forces and U.S. intercontinental systems were adequate deterrence against a Soviet attack.

Moreover, skeptics of the conservative view argued that Soviet naval forces would be severely constrained in wartime because of the limited number of access points to the open ocean available to the four Soviet navies (the Baltic Fleet, the Black Sea Fleet, the Northern Fleet, and the Pacific Fleet) and their inability to perform their several possible naval missions simultaneously. These skeptics noted that all the Soviet gains in the developing world in the 1970s paled in comparison with the major setbacks Soviet diplomacy had suffered in China and Egypt. And, despite the persistent maltreatment of its citizens, Soviet government policies in the late twentieth century were a far cry from the reign of terror that existed from the late 1930s through the early 1950s under Joseph Stalin.

In general, however, the American mood in the early 1980s was marked by considerable wariness of Soviet military capabilities, a deep distrust of the Soviet Union's political intentions, and support for a more assertive American foreign policy than had been practiced since the end of the Vietnam War together with substantial budgetary increases to strengthen the U.S. defense posture. At the same time, Americans wanted to reach equitable nuclear arms control agreements with the Soviet Union where feasible. This was the basis of the Reagan administration's approach to national security policy and nuclear weapons.

Many writers have addressed these issues. A large body of popular and specialized literature exists on the history of nuclear weapons competition and nuclear arms control, on theories of nuclear strategy, and on particular aspects of U.S.-Soviet relations. Yet I believe that in the complex public debate and obscure language used to examine these issues and to defend certain positions, many specialists and the general public have lost sight of the central problems of the nuclear age, problems seemingly without satisfactory solutions. The purpose of this book is to identify these dilemmas and to defend the proposition that because they are inherently unresolvable, they provide the boundaries within which effective policies should be formed and implemented.

Rather than seek to write a comprehensive review of the subject, I have intentionally selected seven boundary conditions for examination in this study.

—The dominant trait of the Russian national character is a pervasive insecurity, which even the top Soviet leaders always feel. Tight domestic political controls and an expansionist but cautious foreign policy are natural concomitants of this insecurity. Maintaining a large and powerful force of nuclear weapons serves both the domestic and international needs of Soviet leaders, who will use this force as a policy instrument as long as the existing Soviet political system endures. Ironically, however, Soviet maintenance of nuclear forces makes other countries, principally the United States, feel insecure, and their reactions only exacerbate Soviet fears and concerns.

—The dominant trait of the American national character is a dual personality, which demands that the United States be both second to none in military power and the principal promoter of world peace. Consequently, Americans support the maintenance of a large nuclear arsenal and at the same time want to abolish these weapons. This split leads to inconsistent policies toward the Soviet Union, which shift from belligerent to accommodating as first the one type and then the other fails to "solve" the outstanding differences between the superpowers.

—Technological progress is leading to the introduction of high-accuracy weapon systems that allow for a more controlled and discriminating nuclear force. Simultaneously, these systems increase the vulnerability to attack of all stationary and many mobile targets, which heightens the sense of insecurity among the leaders and peoples of both the United States and the Soviet Union.

—Both superpowers articulate declaratory policies to enhance the deterrent value of their nuclear forces. Yet the relationship between these policies and actual decisions made during a nuclear war is likely to be highly tenuous. Thus, what we need to say in peacetime is not a useful predictor of what we might wish to do or be capable of doing in wartime.

—Arms control has mistakenly evolved into formal negotiations with the chief goal being limitations on or reductions of total force sizes. This approach, however, has failed to contain or control the principal threats to each side's forces. It is difficult to sustain American political support for arms control if threats are not controlled as well.

—America's European and Japanese allies are caught between entrapment and abandonment by the United States. To satisfy the allies' political and psychological needs, the United States needs to guarantee their security in part through the threatened use of nuclear weapons, even if it is unlikely that the United States would immediately use such weapons in wartime.

—The intensity of the U.S.-Soviet nuclear competition stimulates and legitimizes the process by which other countries acquire nuclear weapons. This process heightens insecurity among regional powers who believe that the likelihood of nuclear weapon use is increasing.

These seven dilemmas each highlight the sense of vulnerability—the capacity to be wounded—that the introduction of nuclear weapons has caused in our time. In four decades we have moved from the testing of the first nuclear device to the accumulation of 60,000 nuclear weapons in the American and Soviet arsenals alone. Yet since World War II no nuclear weapons have been used in anger. A nuclear stalemate between the two powers has long existed. This book, and the dilemmas it addresses, is intended to show how we have arrived at this state of affairs and to evaluate whether prospective developments pose a fundamental threat to the stalemate. In short, I seek to argue that the "nuclear stalemate" has served us well despite its inherent dangers and that we should try to preserve it rather than to undermine it.

CHAPTER TWO

The Absence of Security in the Russian Experience

Change not procedure.
Custom is the soul of states.
Alexander Pushkin, *Boris Godunov*

ALL TOO OFTEN contemporary studies of nuclear weapon policies focus almost exclusively on the evolution of the forces, the doctrines for their use, and the means for controlling them unilaterally or through negotiated agreement. But nuclear weapons are manufactured and deployed by governments, and governments are composed of individuals who function in particular social, economic, and political settings. Surely, therefore, no examination of the influence of nuclear weapons in our time could be fully valid without confronting the nature of the Soviet and American states and the objectives that are satisfied through the acquisition of these weapons. What needs to be explored first, then, are not technical issues but national character, what Samuel Coleridge described as "an invisible spirit that breathes through a whole people, and is participated in by all, though not by all alike; a spirit which gives a color and character both to their virtues and vices."[1]

For the Soviet Union this is easier said than done. It was on October 1, 1939—two weeks after Soviet troops had crossed the Polish frontier and rushed to the Vistula River to help Nazi Germany dismember Poland—that Winston Churchill, not yet British prime minister, offered the now famous observation that Russian behavior cannot be forecast because "it is a riddle wrapped in a mystery inside an enigma." Churchill

1. As quoted in Hans Morgenthau, *Politics among Nations,* 3d ed. (Alfred A. Knopf, 1964), p. 126. Also useful in this regard are several works on "strategic culture." See Ken Booth, *Strategy and Ethnocentrism* (Holmes and Meier, 1979), and Jack L. Snyder, *The Soviet Strategic Culture: Implications for Limited Nuclear Operations,* R-2154-AF, prepared for the U.S. Air Force (Santa Monica, Calif.: Rand Corp., September 1977).

Table 2-1. *Different Conceptions of Soviet Policy*

Item	*School 1*	*School 2*
Thrust of Soviet foreign policy	Aggressive	Defensive
Dominant Soviet preoccupation	Expansion	Consolidation
Soviet strategy	Grand design	Opportunistic
Assessment of military balances	Emphasizes Soviet strengths, American weaknesses	Emphasizes American strengths, Soviet weaknesses
Role of arms control	Limited	Considerable
Assessment of Soviet outlook	Optimistic	Pessimistic
Appropriate Western response	Promote anti-Soviet united front; strengthen U.S. nuclear and conventional force postures	Accentuate cooperative aspects of relationship; emphasize strengthening U.S. nonnuclear forces

went on to say, "But perhaps there is a key. That key is Russian national interest."[2]

Since World War II, as relations with the Soviet Union came to dominate American foreign policy, many observers have tried to unwrap the riddle of Russian behavior by seeing, as nearly as is possible, the Soviet national interest as the Soviet leadership sees it rather than through the lens of Western experiences and American judgments. Nonetheless, Americans have been unable to agree among themselves about why the Soviet Union pursues certain policies or how best to respond to these policies.

Those who claim influence on U.S. policy differ on the major points summarized in table 2-1. Members of school 1 see a Soviet Union whose foreign policy is fundamentally in conflict with the interests of capitalist countries. According to this view, the Soviet leadership primarily wants to expand Soviet influence throughout the world, encourage the proliferation of communist governments that support Soviet policies, and move irrevocably toward the demise of capitalism.

To achieve these goals, according to this view, the Soviets have

2. Excerpts from Churchill's remarks, aired by the British Broadcasting Corporation, can be found in Winston Churchill, *The Second World War: The Gathering Storm* (Bantam Books, 1961), p. 399.

adhered to a grand strategy that has evolved through different phases since World War II. The first phase concentrated on expanding Soviet influence in Europe, capitalizing on the position of Soviet military forces at the end of World War II and on the effectiveness of procommunist groups within individual European states. Eventually thwarted in Europe through the establishment of NATO and the economic recovery of Western European nations, in the 1960s the Soviets shifted their attention to wars of national liberation in third world areas, particularly in Southeast Asia. Now, having achieved victory in this region, the Soviets have turned to the Persian Gulf and, with the help of Cuba, to sub-Saharan Africa and Latin America. Having seen how vulnerable the West is to the oil pricing and supply policies of the Organization of Petroleum Exporting Countries (OPEC), the Soviet Union is now trying to create instability from Afghanistan to South Yemen to ultimately gain control over the flow of oil that is essential for the maintenance of the industrialized economies. Moreover, because of the Cuban forces and Soviet arms being used to instigate revolution in Latin America, U.S. forces are likely to be committed to conflicts in the West, thereby reducing the effectiveness of American counterpressure against Soviet expansionism nearer Soviet borders.

At the same time, the Soviets have invested enormously in surpassing the United States as the world's foremost military power. The Soviet Union has now achieved superiority in most of the static indicators of the strategic nuclear forces, most significantly in the countermilitary potential of its weapons. Soviet superiority over NATO in the European theater continues to grow, particularly with the deployment of mobile intermediate-range ballistic missiles and modernized conventional forces. The Soviets now have a global naval force, with growing capability to project military and political power at great distances from the borders of the USSR. The increasing vulnerability to attack of the U.S. land-based missile force and the weaknesses within NATO present attractive opportunities for Soviet adventurism. If present trends continue, the Soviets will take advantage of their military position to coerce the West for political and economic gain. In a future equivalent to the Cuban missile crisis, it is the Soviets who would prevail. In this political-military context, arms control negotiations with the Soviets are of limited or negative value, since ensuing agreements would either accentuate Soviet advantages or inhibit the United States from taking the unilateral actions necessary to restore strategic stability.

The outlook of the Soviet leadership is, consequently, optimistic, at least in the short term. The most appropriate American response to this state of affairs is to redress immediately the military imbalance in both nuclear and conventional forces and to establish a diplomatically united front against the Soviet Union, joining the countries in NATO, Japan, and perhaps China. The purpose of this front would be to complicate Soviet defense planning and, over time, to show the Soviet leadership that victory over the West is not feasible.

Members of school 2 (see table 2-1) see a very different picture. They see a Soviet foreign policy dominated by extreme caution marked by defensive opportunism. According to this school of thought, maintenance of the Soviet communist regime and of friendly states contiguous to the Soviet Union, especially in Eastern Europe, remains the highest priority of the ruling elite, with the goal of "victory over capitalism" mere rhetoric that lost its operational significance long ago. This elite, it is argued, continues to be most influenced by the horrors of World War II, and its main concern is to consolidate the gains made since 1945 to ensure that the Soviet homeland is never again ravaged by war.

These achievements alone will be difficult to sustain. Economic problems—marginal agricultural productivity, skilled-labor shortages, declines in the growth of industrial output—are seen as chronic and insoluble without either major economic transfusions from the West (now unlikely) or a significant reorganization of the entire economic system (politically highly dangerous). Geopolitically, there are major trouble spots, especially in Poland and Afghanistan and with China.

Proponents of this school of thought see the Soviet Union—beset by these difficulties—pursuing an opportunistic foreign policy divorced from any grand strategy, a highly pragmatic and cautious policy that is producing mixed results. For every victory in Angola, Ethiopia, or Mozambique, there is defeat in Egypt, Ghana, or Indonesia.

Although the Soviets clearly have increased their military power, the Soviet leadership remains deterred from initiating acts that could trigger a Soviet-American conflict. According to this view, the Soviet military buildup is a result more of Russian paranoia and bureaucratic inertia than of aggressive intent. Moreover, too many American analysts are unduly pessimistic about American military strength in relation to Soviet forces, according to members of school 2. The most effective public relations efforts supporting Soviet military capability and its political utility are those mounted not by the Soviets but by the conservative members of school 1. In fact, the Soviets can gain little or no political advantage from their military forces unless the United States permits them to do so, and the Soviets know full well that a Warsaw Pact attack on NATO forces could lead to nuclear war. No rational Soviet leader would prefer such a contingency to the present one, say members of school 2. Negotiated arms control agreements, therefore, could stabilize the Soviet-American nuclear balance, a highly desirable alternative to more intense arms competition.

Faced with a dynamic Western capitalist system with an enormous technological advantage over the Soviet Union, Soviet leaders are fundamentally pessimistic about the long-term abilities of the communist system to compete effectively in economic or political terms, according to the second school of thought. The next generation of Soviet leaders, better educated and more widely traveled in the West than their predecessors, must become convinced that the Soviet-American relationship can enhance cooperation and reduce confrontation. The West should, therefore, provide incentives to bolster cooperation and should avoid economic, political, and military initiatives designed to isolate the Soviet Union.

These two schools of thought are obviously oversimplified, and readers might find that they share some judgments from each school. But what is the basis for these positions? How do we know what we think we know about the Soviet Union? Even talented individuals who have devoted their professional lives to the study of Soviet behavior emphasize different characteristics and reach widely varying conclusions.

Alexander Solzhenitsyn, himself a victim of Soviet repression, concludes that it is the communist system and the individuals who control it that determine the innate aggressiveness of the government. He asserts:

> Never has the Politburo numbered a humane or peace-loving man among its members. The Communist bureaucracy is not constituted to allow men of that caliber to rise to the top—they would instantly suffocate there. . . . Communism will never be halted by negotiations or through the machinations of détente. It can be halted only by force from without or by disintegration from within.[3]

Richard Pipes, a student of Russian history and a former senior adviser on Soviet affairs on the Reagan administration's National Security Council, sees Soviet aggressiveness more as a consequence of

3. Aleksandr Solzhenitsyn, "Misconceptions about Russia Are a Threat to America," *Foreign Affairs,* vol. 58 (Spring 1980), pp. 807, 833–34.

a snapshot of a vast and complex land. Yet a century later, Walter Bedell Smith, U.S. ambassador to the Soviet Union, was struck by the relevance of Custine's remarks to the country in which he himself was serving. Smith observed, "I could have taken many pages verbatim from his journal and, after substituting present-day names and dates for those of a century ago, have sent them to the State Department as my own official reports."[10]

Pre-Soviet Russia had various experiences with reform and even flirtations with democracy. As far back as the twelfth century, for example, Russian principalities relied on a political organization that included a city assembly known as the *veche*, in which unanimous consent of all adult males was required before decisions could be adopted. During the Mongol occupation of Russia in the thirteenth century, St. Sergius of Radonezh, founder of a monastery near Moscow and an influential pioneer in the Christianization of the rural population, stressed humaneness and the importance of individual initiative. Alexander I, who reigned in the first quarter of the nineteenth century, was known for his broad, international perspective and engaged in correspondence with Thomas Jefferson, then president of the United States, about American principles of government that he intended to adopt in reforming Russia's autocracy.

In March 1861, under the reign of Alexander II, serfdom was officially abolished. Peasants became free men, although they were not able either to become individual property owners or to acquire their full civil rights, since each peasant village received land only for communal ownership. Constitutional reforms were also implemented in 1905, when Nicholas II granted to the Russian nation fundamental civil liberties, including freedom of speech, assembly, and organization and guaranteed that no law could be enacted without the consent of a newly created house of representatives, the Duma.

These are just a few of the Russian efforts at reform. But they did not dominate Russian history; they were the exceptions rather than the rule. In fact, the central tendencies that most powerfully describe the Russian traditions inherited by Lenin can be found in the Marquis de Custine's memoirs and can be summarized under the following headings.

10. See Smith's introduction to ibid., pp. 8–9. General Smith, who negotiated and accepted for the Allies in World War II the surrender of Italy in 1943 and Germany in 1945, served as ambassador to Moscow from 1946 to 1949.

Hierarchy. Russian society has historically been a mixture of the dominant and the dominated. For 400 years the czar ruled with an iron hand and a God-given right. Even in the early twentieth century, when he no longer had an iron hand, he was convinced he still possessed God's sanction. This sense of hierarchy was not restricted to the relationship between the czar and his subjects; it permeated all levels of society. The aristocrats frowned upon the bourgeoisie, which grew rapidly after the mid-nineteenth century. The landed gentry dominated the peasants. The Russians as an ethnic group developed a strong sense of superiority and intolerance toward the non-Russian minorities with whom they shared a common state—the Ukrainians, the Belorussians, the Uzbeks, the Georgians, the Kazakhs, and the like. Whereas sovereignty became a well-understood concept when applied to government, the Jeffersonian notion of sovereignty of the people never took hold. Russians never accepted the view that the purpose of government was to serve the people. Instead, they believed, it was the duty of the people to serve the state.

Theocratic co-optation. Despite occasional liberalizing policies that the Russian Orthodox church initiated, it basically reinforced the rule of the aristocracy rather than modified or undermined it. The Orthodox church was derivative of Byzantium, not the Christianity of Rome or Luther. It was steeped in ritual, mysticism, and dogmatic complexities and began to be eclipsed by the state as early as the seventeenth century, when a special government department was established under Peter the Great to supervise church administration. Indeed, the church continued under the guardianship of the government until 1917. Consequently, the role that various theologians and church groups played in the development of European and American liberalism was largely absent from the Russian experience.

Sense of inferiority. From the time Russian leaders first encountered Western Europeans, they were preoccupied and at times obsessed by their relative inferiority. By virtually all standards—military strength, economic output, managerial efficiency, language, custom, and dress—the Russian elite, from the time of Peter the Great, felt they had much to learn and to take from European society. Understandably, there was for centuries tension within Russia about how much the society should be built on its indigenous Slavic institutions and how much it should emulate the West. The Russian attitude was never simply one of pure admiration. It was always a complex mixture of envy and scorn, reflecting more a desire to take from Europe what Russia needed rather than to adhere to a full-scale Russian remodeling in the European image.[11] The Bolsheviks' predecessors never could sustain a consistent long-term policy toward Europe. Instead they attempted emulation, isolation, or domination.

Deep-seated nationalism. Love of the land and adoration for the national entity of Mother Russia were consistent bulwarks of the Russian character. An intense, almost mystical attachment to the motherland was pervasive among all socioeconomic classes despite the harshness of life for most citizens. Curiously, Russians were able to maintain their unshakable loyalty to the nation-state and simultaneously to be critical or pessimistic about daily life under the czar. These dual sentiments recur persistently in the great Russian literature and music of the nineteenth century. Despite the diversity of settings and emphases in these works, Alexander Pushkin's *Boris Godunov,* Fyodor Dostoevsky's *Crime and Punishment,* Leo Tolstoy's *War and Peace,* Anton Chekhov's *The Cherry Orchard,* Peter Ilich Tchaikovsky's *Sixth Symphony,* and many less well known works portray the depths of feeling and despair in the individual Russian existence as well as the strength and perseverance of the Russian nation.

Territorial expansion despite military weakness. Geographically, Russia expanded between the seventeenth and nineteenth centuries along each of the four distinct topographical zones that run west-to-east across the Eurasian land mass: the frozen tundra in the north, the great forests in the north-central region, the steppes or plains in the central and south-central region, and the deserts in the south. (Traditionally, the Russians saw the Ural Mountains as a bridge between European and Asian Russia instead of as a barrier between the two regions, as Western Europeans have often contended it was.)

But the price of territorial expansion, which was always contiguous to territory already conquered rather than in the form of overseas colonies, was high. Russia consolidated its western boundaries, but only temporarily, after bloody struggles with Sweden, Poland, and Prussia and after the Napoleonic invasion of 1812 had been repulsed. In the south, Russian territorial ambitions triggered several conflicts with

11. This approach is reflected in the analyses of V. O. Kluchevsky, the preeminent Russian historian of the nineteenth century. Kluchevsky claims that Peter the Great once stated, "For a few score more years only shall we need Europe. Then shall we be able to turn our backs upon her." See V. O. Kluchevsky, *A History of Russia*, vol. 4 (Russell and Russell, 1960), pp. 220–24.

Turkey, culminating in a major Russian defeat in the Crimean War of 1854–56, in which Turkey was supported by Britain and France. The loss of the right to maintain a fleet in the Black Sea was one of several costs of that war for Russia. In the Far East, earlier Russian gains in the acquisition of Chinese ports and other territories were more than offset by Russia's humiliating defeat in the Russo-Japanese War of 1904–05, the first time since the era of the Mongols that a European power had been defeated by an Asian people.

Finally, in World War I, the full extent of Russia's military weakness was exposed. In 1914 Russia entered the war against Germany and Austria-Hungary without in any way being adequately prepared. Nearly 4 million Russian soldiers were killed, wounded, or captured in the first ten months of fighting. By 1915 as many as 25 percent of Russian soldiers sent to the front were unarmed; their only weapons were the rifles of their own fallen comrades. It was only after the Bolsheviks seized power that Russia could extricate itself from the war through negotiations with Germany. The Treaty of Brest-Litovsk, signed in March 1918, codified the Russian loss to Germany. Lenin and his colleagues had to yield the Ukraine, Latvia, Lithuania, and Estonia to the Germans as well as part of the Transcaucasus to the Turks.

In sum, territorial expansion did not bring Russia any sense of stability in its international relations. Rather, it prompted increased commitments, exacerbated threats, exposed military weaknesses, and often culminated in disastrous and humiliating defeats.

Foreign defeats and domestic change. Changes in leadership in Russia followed certain established patterns. First, changes were infrequent. Several czars (Ivan III, Ivan the Terrible, and Peter the Great) each reigned for more than forty years. From the beginning of the nineteenth century to the Bolshevik Revolution, only five czars ruled Russia, with twenty-five years being the average reign. Second, change emanated from the top. Compared with Europe, Russia experienced far fewer mass uprisings and widespread expressions of discontent. Instead, leadership changes resulted from struggles among the elite—in the earlier days among rivals within the aristocracy and in the last days of pre-Soviet Russia among the relatively small numbers from the intelligentsia and the working classes. Characteristically, not many Russians were politically aware and very few were politically active. Third, domestic reform and a change in leadership were often consequences of foreign disasters. The abolition of serfdom and other reforms adopted in the early 1860s were an immediate result of pressures unleashed following Russia's defeat in the Crimean War. The constitutional innovations of 1905 resulted directly from the czar's weakened authority following Russia's defeat at the hands of the Japanese. And, of course, it was Russia's calamitous experience in World War I that led to the czar's abdication and the Bolsheviks' ultimate victory.

Secrecy and terror. The imposition of secrecy and terror by the ruling elite on the mass population was endemic to Russian society throughout the czarist period. Despite sporadic episodes of reform, Russia's rulers routinely governed in secret, shielded their subjects from knowledge of the outside world and from physical access to it, and applied the most extreme forms of mental and physical abuse to those who dissented openly from official policies.[12] Even in the early twentieth century, the press was viewed as an arm of government and communicated ideas, decisions, and interpretations of events consistent with formally approved policies. The ruling elite judged publications expressing independent views as purely adversarial to the state. The tradition of a free, responsible, but critical press as the "fourth estate" of government—a central element in Western society—never developed in Russia. Even during the operation of the Duma, from 1905 to 1917, when for the first time in Russian history political parties formed openly, they functioned as elite groups that based their differing programs for social, political, and economic reform on theoretical tracts that had little to do with the grievances of most Russian citizens. Moreover, the entire Russian legal structure, especially the police and the other elements of the criminal justice system, was designed to serve the interests of the state rather than to protect the rights of citizens. In this context, secrecy and terror served two interrelated czarist goals: they squelched internal dissent and discouraged receipt of information from abroad that would have given Russians more facts by which to judge the inadequacies of their own society, both in absolute terms and in relation to that of Western Europe. Secrecy and terror, therefore, were essential for preserving the autocracy.

12. It is interesting to note, for example, in light of contemporary Soviet government approaches to the silencing of dissidents, that Peter Chaadaev (an officer of the guards during the reign of Nicholas I) was thought to be too critical of Slavic institutions in his endorsement of Russian adoption of European practices. Consequently, Chaadaev was declared legally insane. His essays were published in a work called *Apology of a Madman.*

There is no doubt that in the fifty years or so between the freeing of the serfs and the Bolshevik Revolution, several liberalizing reforms took hold in Russia. But overall, it was these innate characteristics—hierarchy, theocratic co-optation, a sense of inferiority, deep-seated nationalism, territorial expansion despite military weakness, foreign defeats and domestic change, and secrecy and terror—that formed the foundation upon which Lenin and his colleagues began building the socialist order.

The Soviet Achievement

In the more than sixty years since the Bolshevik Revolution, Soviet achievements have been nothing short of colossal, as have the costs. Joseph Stalin, having consolidated power by 1928, sought to remedy what he saw as the fundamental weakness of czarist Russia. This was made abundantly clear in his defense of the first Five-Year Plan, given in a now famous statement before the First All-Union Conference of Managers of Socialist Industry in February 1931:

> To slacken the tempo would mean falling behind. And those who fall behind get beaten. But we do not want to be beaten. No, we refuse to be beaten! One feature of the history of old Russia was the continual beatings she suffered for falling behind, for her backwardness. She was beaten by the Mongol khans. She was beaten by the Turkish beys. She was beaten by the Swedish feudal lords. She was beaten by the Polish and Lithuanian gentry. She was beaten by the British and French capitalists. She was beaten by the Japanese barons. All beat her—for her backwardness: for military backwardness, for cultural backwardness, for political backwardness, for industrial backwardness, for agricultural backwardness. . . . It is the jungle law of capitalism. You are backward, you are weak—therefore you are wrong; hence, you can be beaten and enslaved. You are mighty—therefore you are right; hence, we must be wary of you. That is why we must no longer lag behind. . . . We are fifty or a hundred years behind the advanced countries. We must make good this distance in ten years. Either we do it, or they crush us.[13]

Thus the leader of a nation-state that covered roughly one-sixth of the earth's land mass recounted its successive military defeats, thereby playing to the insecurities of his audience to motivate them to implement his plans.

13. J. Stalin, *Problems of Leninism* (Moscow: Foreign Languages Publishing House, 1947), p. 356.

Throughout the 1930s Stalin sought primarily to strengthen his grip on all levers of power in the Soviet system while building his country's economic and military might; a distinctly secondary consideration was the spreading of communism abroad. In satisfying his primary goals, Stalin invoked the harshest measures imaginable, often citing the ideas and writings of Lenin to justify his acts. Perhaps motivated partly by his own personal tragedy,[14] Stalin struck out against all his potential rivals, real and imagined, in both the Communist party and the Red Army. By the end of 1938, Stalin had killed virtually all the party's top leaders who had survived the revolution. By Nikita Khrushchev's account, of the 139 members and candidates of the party's Central Committee who were elected at the Seventeenth Party Congress held in 1934, 98 people, or 70 percent, were arrested and shot.[15] The army's entire top leadership and key figures in the secret police, diplomatic service, and trade unions all were executed. The total number of people killed in what came to be known as the Great Purge has never been documented; Western estimates range from tens of thousands to several million. In addition, the harshness of rapid industrialization and agricultural collectivization also took an enormous toll in human lives from among those who resisted the policies or who were too weak to carry them out.

As Khrushchev later observed, all those around Stalin were "temporary people" and, once Stalin became distrustful, "it would be your turn to follow those who were no longer among the living."[16] Stalin, Khrushchev emphasized, ran Soviet foreign policy single-handedly. He alone made all the major decisions—to enter into the neutrality pact with Nazi Germany in 1939 (to gain time to prepare for a possible German invasion of the Soviet Union and to take advantage of the pact's secret protocol that called for a Russo-German division of Poland); to ignore warnings about Operation Barbarossa, Adolf Hitler's invasion of Russia in June 1941; to ask the allies to open a second front against Hitler to relieve German pressure on the Red Army; and to orchestrate wartime diplomacy with Prime Minister Winston Chruchill and President Franklin

14. Stalin's second wife committed suicide in 1932 following a brief argument between them, and Stalin immediately afterward adopted recluse-like behavior. See the account by his daughter, Svetlana Alliluyeva, *Twenty Letters to a Friend* (Harper and Row, 1967), pp. 112–13, 205.

15. Cited in Khrushchev's famous "de-Stalinization" speech at the Twentieth Party Congress in 1956. See Strobe Talbott, ed. and trans., *Khrushchev Remembers* (Little, Brown, 1970), p. 572.

16. Ibid., p. 307.

D. Roosevelt so that the Soviet Union would be in the most favorable military position at the war's end to control the peace that would follow.

The Soviet performance in World War II illustrated all the strengths and weaknesses of the Russian character. In this, the Great Patriotic War, the Russian people and the Red Army exhibited unbelievable tenacity, an almost superhuman ability to prevail in the face of catastrophic circumstances. Despite the death and destruction inflicted by Nazi forces on Soviet Russia, the experience for the nation as a whole was enormously positive. Most significantly, the war demonstrated a cohesiveness of purpose between the leadership and the people that was virtually unprecedented in Russian history, initial Ukrainian pro-German support to the contrary notwithstanding. The very survival of the nation rested on the shoulders of millions of Soviet citizens, all united in the common goal of crushing the hated German invaders.

The war revealed many Soviet weaknesses as well. Soviet military preparedness was substantially deficient.[17] Also, because of Stalin's purges in the late 1930s, the Red Army was led by an inexperienced and initially inept officer corps that had to acquire on-the-job training under extraordinarily adverse conditions.

Nonetheless, over time the Red Army performed with great distinction. Innovative Soviet tactics were devised and adopted, as in the battle of Stalingrad; these proved decisive against a sophisticated and experienced German army. The Russians also eventually manufactured and introduced in combat, with notable effectiveness, military equipment

17. Since the war Soviet military figures have persistently reviewed these deficiencies and the lessons to be learned from them. As one example, consider these observations by N. Ogarkov, then chief of the armed forces general staff and first deputy defense minister: "Back in the prewar years our military science, far outstripping bourgeois military thought, elaborated the advanced theory of the in-depth [*glubokaya*] operation—a fundamentally new method of conducting active offensive operations using massed, technically equipped armies. In accordance with this theory in 1932 our country became the first in the world to create major formations of tank troops in the form of mechanized corps. By 1936 four such corps had already been formed, and these were later reformed as tank corps. However, subsequently, because of a number of objective and subjective factors, incorrect conclusions were drawn, based solely on limited experience of the use of tanks in Spain. As a result in 1939 the corps which had been created were disbanded, and it was again proposed to use cavalry as the exploitation echelon in operations. This tenet was subsequently corrected, and by 1942 we had created not only tank corps, but tank armies, though it would have been better to have had them before the start of the war." See N. Ogarkov, "Guarding Peaceful Labor," *Kommunist*, no. 10 (July 1981), pp. 80–91, reprinted in Joint Publications Research Service, *Translations from Kommunist*, JPRS 79074, September 25, 1981, p. 92.

superior to that of the enemy (for example, the T-34 tank, much more advanced than the German panzer). Most significantly, the Red Army, Stalin, Soviet communism, and the Russian people were victorious. And no matter what the cost of victory, victory itself was of incalculable value.

The Soviets seemed to have learned three fundamental lessons from the war. First, it demonstrated the strength of the Russian people and the communist system, even if, objectively, the communist system had little to do with the Russian victory. Under Stalin and his successors, every effort has been made—through television, film, the press, historical "scholarship," and all other forms of communication at the government's disposal—to emphasize the close connection between triumph in the war and the legitimacy and effectiveness of communism. Indeed, sophisticated and unrelenting references to both the war and the promise that daily Soviet life will consistently improve have been the two main inducements the Soviet leadership has used to cement loyalty to the communist system.

Second, the war confirmed the requirement to strive to acquire the military capability to overpower all potential enemies and the need to develop and sustain the industrial base essential to this capability. Looking at the war not only from a pragmatic perspective but through the lens of communist ideology and dogma, Soviet leaders were more convinced than ever of basic principles articulated by Friedrich Engels about seventy years earlier:

> Always and everywhere it is the economic conditions and instruments of force which help "force" to victory, and without these, force ceases to be force. And anyone who tried to reform methods of warfare from the opposite standpoint . . . would certainly reap nothing but a beating.[18]

In short, the war defined the economic priority for the postwar period: to establish and modernize an industrial base dedicated first and foremost to the production of war matériel. This priority was not established solely to ensure Soviet security; it was also adopted to preserve the regime. For what must have been uppermost among Stalin's concerns after the war was the immutable logic that had plagued his czarist predecessors since the mid-nineteenth century: industrial weakness,

18. Friedrich Engels, *Anti-Dühring* (Berlin: Dietz Publishing House, 1970), pp. 185, 190. These views have been repeated consistently since the war. For example, they are paraphrased in *Khrushchev Remembers,* p. 158, and in the article by Marshal Ogarkov cited above, note 17.

reflected in military inferiority, produced foreign policy defeats that led to pressure for domestic political and economic reform that could trigger a regime change from the top. The surest way to strengthen his hold on the leadership was to rebuild Soviet industrial and military might.

The final lesson of the war was that, if at all possible, the states contiguous to the Soviet Union should not be permitted to threaten Soviet security. Indeed, Stalin's great achievements in foreign policy were to solidify Soviet control over the countries in Eastern Europe (Bulgaria as well as the four states contiguous to the Soviet Union); guarantee the long-term division of Germany with a powerful Soviet military presence in the eastern third of the country; bring the Soviet Union into the nuclear age as a nuclear power; and, in the process, establish it as one of the two superpowers in the postwar period. It has been the task of Stalin's successors to preserve and build on his foreign policy successes.

Contemporary Soviet Characteristics

Although communism has been deeply entrenched in the Soviet Union for more than six decades, and its fundamental tenets are the antithesis of the autocracy it overthrew, the similarities between czarist and Soviet Russia are as striking as the differences. Hierarchy remains central to Russian life, although now it is the Communist party that controls society rather than the czar and his agents. The party is an elite organization that has doubled in size since Khrushchev's days, from roughly 7 million to 15 million members, but it has yet to exceed 7 percent of the total population.[19]

Through an elaborate system of controls and procedures the party dominates all phases of Soviet life. All institutions in the country—factories, farms, schools, stores, theaters—report through a complex, centralized bureaucratic chain of command to a ministry in Moscow whose authority is ultimately determined by the party. The party's fundamental principles are emphasized and reemphasized through the media. The sole path to power in the Soviet system is to climb the ranks of the party structure, through various local, regional, and functional

19. Statistics on party membership and subsequent citations concerning army-party relations draw on National Foreign Assessment Center, *Political Control of the Soviet Armed Forces* (Washington, D.C.: Central Intelligence Agency, 1980), pp. 1–11.

bodies to the party's Central Committee (which had 319 voting members when the Twenty-sixth Party Congress was convened in February 1981) and finally to the Politburo (which had eleven voting members in 1984).

To ensure that the armed forces are unequivocally subordinate to party directives, the Soviet regime maintains the Main Political Directorate (MPD) of the Soviet army and navy, which extends from the Defense Ministry in Moscow to the company level in the field. The MPD both explains party policies to the troops and receives political feedback from all levels of command. It verifies that party orders are implemented throughout the military establishment. It participates in the selection, assignment, and political evaluation of military officers. It supervises the content of the military press. It controls the training, research, and curriculum development of social science material in all military academies and educational institutions. Political control of the Soviet armed forces would appear to be pervasive and effective.

The czars' efforts at theocratic co-optation have been replaced by the official endorsement of atheism and the outlawing of formal religious practices. The communist regime has always been somewhat more tolerant of expressions of religious belief by members of both the Russian Orthodox church and the Islamic faith than its official posture suggests.[20] But this tolerance has most often been exercised for the explicit political purpose of pacifying non-Russian ethnic groups. The entire Soviet structure of incentives is geared toward the observance of communist ideology and to no other beliefs.

Despite the extraordinary power and achievement of the Soviet system, an endemic sense of inferiority persists. It is revealed in many forms. Consider this example of Russian humor from the Khrushchev period:

> An advisor came in to Premier Khrushchev very exultantly one day. "I can prove that Adam and Eve were Russian," he said. "We can now claim the invention of man along with our invention of radio, automobiles, telephone and television." Khrushchev asked, "How? You know the United States likes to ridicule our claims in such matters and we must be very careful that we can document it." "I am completely confident," said the official. "Adam and Eve had no clothes, they had no roof over their heads and had only apples. Yet they thought they lived in Paradise. What else could they have been but Russians?"[21]

20. The Soviet Union has an estimated population of 30 million Muslims. This is the ninth largest concentration of Muslims in the world, after Indonesia, China, Bangladesh, Pakistan, India, Turkey, Egypt, and Iran.

21. *USSR Humor*, compiled by Charles Winick (Mount Vernon, N.Y.: Peter Pauper Press, 1964), p. 17.

The anecdote lampoons several characteristics long evident in Russian culture: the strong desire to be superior to the West and particularly to the United States; the realization of the inadequacies of the society; and the willingness to distort or falsify reality to maintain the illusion of Soviet superiority, a willingness motivated by a deep sense of national pride. These traits are peculiarly Russian; it is hard to conceive of such a tale being told by the Americans, British, French, Germans, Japanese, or Chinese.

The Soviet economy justifies this sense of inferiority. Although the total gross national product (GNP) of the Soviet Union has for some time been roughly half that of the United States, causing the Soviet Union to rank second in the world (though soon to be overtaken by Japan), most other measures of economic performance indicate chronic difficulties. Economic growth rates have declined to very modest levels (for example, a less than 3 percent average annual increase in both GNP and in real volume of investment projected for 1981–85), explained in part by managerial inefficiencies and low worker productivity. High absenteeism and widespread alcoholism are two of the interrelated causes for this poor economic performance.

Agricultural shortfalls, highly dependent on climatic conditions, were particularly severe in the late 1970s and early 1980s. Two extremely adverse consequences resulted: the need to depend heavily on imports from the West to feed the Soviet people—itself a politically humiliating development—and the depletion of scarce hard-currency reserves to pay for these imports. (The currency was obtained in part through the sale of large quantities of gold in the international money markets.) Moreover, in the face of rising energy consumption, glaring weaknesses have been revealed in Soviet distribution systems, managerial practices, and technological capabilities. These weaknesses have impeded the transfer of energy produced from the proven reserves east of the Ural Mountains, the location of 80 percent of fossil-fuel reserves, to the citizens and industries west of the Urals, where 75 percent of the energy produced is consumed.

With an official GNP per capita less than that of Czechoslovakia, East Germany, and the eighteen Western industrial market economies, Soviet citizens have been forced to establish a large "second economy" through black market and other underground activities. Beyond the reach or with the acquiescence of the official apparatus, this second economy enables Russians to acquire additional goods and services, and is operated alongside an officially approved economic reward structure created for the elite. This latter structure gives privileged citizens access to the best Western goods—the more privileged the citizen, the better the goods—and is in sharp contrast to the official commitment to a truly egalitarian society. The Soviet Union endures these economic shortcomings while the leadership allocates at least 11 to 13 percent of its annual GNP to defense spending. This allocation increases in real terms at an annual rate of 2 to 4 percent each year, with the defense sector claiming the best talent and most modern facilities that the Soviet system can produce.[22]

The Soviet Union has significantly modified its system of internal repression in the last three decades. Stalin's tactics—politically rigged trials, murders carried out by the secret police, imprisonment and subsequent execution of tens of thousands of individuals in work camps—have been substantially overhauled. Citizens no longer fear arbitrary execution as they did under Stalin. Yet contemporary Soviet society remains gripped by secrecy and repression, by the absence of any toleration for dissent from official government and party policies and regulations. If Solzhenitsyn's description of the Gulag Archipelago is accurate, a system of prison camps, secret police installations, and espionage organizations honeycombs the countryside. Critics of the system or those who openly express the desire to emigrate risk losing their jobs, being ostracized from their professional colleagues and friends, and being harassed and intimidated by state security officials. Persistent criticism could lead to harsher measures: imprisonment in mental institutions, prisons, or work camps or internal exile to places off limits to foreigners. Some critics, seen as embarrassments or threats

22. Particularly useful data on the performance of the Soviet economy are contained in *Allocation of Resources in the Soviet Union and China—1980,* Hearings before the Subcommittee on Priorities and Economy in Government of the Joint Economic Committee, 96 Cong. 2 sess. (Government Printing Office, 1981), pt. 6, especially pp. 135–59. These hearings, conducted annually, are an invaluable source of information on the Soviet economy. See also Alec Nove, "The Economic Problems of Brezhnev's Successors," *Journal of International Affairs,* vol. 32 (Fall–Winter 1978), pp. 201–09; Jeremy Russell, "Energy in the Soviet Union: Problems for Comecon?" *Journal of World Economy,* vol. 4 (September 1981), pp. 291–313; *World Development Report, 1981* (New York: Oxford University Press, 1981), pp. 134–35; and Abram Bergson, "Soviet Economic Slowdown and the 1981–85 Plan," *Problems of Communism,* vol. 30 (May–June 1981), pp. 24–36. A comprehensive critique of the Soviet economy is Marshall I. Goldman, *The U.S.S.R. in Crisis: The Failure of an Economic System* (Norton, 1983). As for the Soviet defense burden, CIA estimates of 3 to 4 percent annual growth after inflation have been revised downward to 2 percent. See "CIA Analysts Now Said to Find U.S. Overstated Soviet Arms Rise," *New York Times,* March 3, 1983.

to the regime because of their international stature and links to the West, are stripped of their citizenship and expelled from the country against their will. In short, although the threat of arbitrary execution no longer exists, the ruling elite uses a variety of coercive measures to retain its monopoly of power.[23]

The Russian Character and the Role of Nuclear Weapons

Individuals in a political structure generally derive their definition of steady-state conditions and of preferential relationships and modes of behavior from many sources: the history of their state and the lessons drawn from this history; the behavior and ideas of great national leaders from the past; the nation's geographic location; the beliefs the people hold about their own culture and its relationship to others; the nature and perceived effectiveness of the political institutions that govern life in the society and the political ideology that has formed those institutions; and the economic structure and its system of rewards and penalties. In all categories the Russian experience is marked not by concepts of equilibrium, stability, balance, or parity. Rather, the essence of the Soviet experience has been dominance, hierarchy, insecurity, and disequilibrium.

The Russians' sense of their history is that they have been weak and

23. Students of Soviet demography and domestic politics point out that during the 1980s the Russians will become a minority in their own country. It has been estimated that there are perhaps 100 million Soviet citizens who have limited or no knowledge of the Russian language. Given the higher birthrates of non-European Soviet citizens, this "nationality problem" is certain to worsen. It is unclear, however, the extent to which the changing ethnic composition of the Soviet citizenry will produce pressure for political change, nor is it clear how coercive techniques of the leadership will be altered to meet this nationality challenge, should it occur. One prominent student of the Soviet political system asserts that "the multinational character of the Soviet Union poses potentially the most serious threat to the legitimacy of the Soviet state and to the stability of the Soviet regime." But in posing the question "why the nationality problem has not become a real nationality crisis," he responds, "we do not have a full answer to this question, and I have not found an answer to it in anything I have read on the subject written in the East or West." See Seweryn Bialer, *Stalin's Successors: Leadership, Stability, and Change in the Soviet Union* (Cambridge University Press, 1980), p. 212. A summary of trends in Soviet demography and their implications for the Soviet system is contained in chapter 10, "Soviet Stability and the National Problem," pp. 207–25. The overriding fact, however, is that, from all available evidence, no significant ideology of opposition has yet materialized to challenge the primacy of the Communist party of the Soviet Union.

have suffered greatly because of this weakness. They can point to few great heroes, especially since the establishment of the Soviet state, and those they revere—Marx, Lenin, Stalin (despite Khrushchev's denunciations)—symbolize through word or deed or both the importance of strength and dominance of the party and the state rather than, for example, equality, compassion, or generosity. Indeed, many who performed notably in service of the state, including Leon Trotsky (extolled by Lenin for his "exceptional abilities"), Viacheslav Molotov (who served as foreign minister from 1939 to 1956 with a single brief interruption), Georgy Zhukov (one of the great military heroes of World War II and subsequent defense minister), and Khrushchev all suffered ignominious falls from power.[24]

In geopolitical terms, the vastness of the borders that the Soviet state must defend has been as great a source of insecurity as the vastness of the state itself has been a source of strength and admiration. Culturally, despite great Russian nationalism, there persists a deep sense of inferiority about the seemingly innate inefficiency and crudeness of the Russian character in contrast to the Western and especially American personality.

The political system and the ideology that supports it stress hierarchy in the extreme, and the entire society is built around a totalitarian system dominated by the Communist party. The party uses a variety of coercive and repressive techniques to suppress diversity and potential opposition and yet is riddled with inner tensions stemming from the lack of any legitimate, established process of governmental succession. The economic system continues to exhibit chronic weaknesses—Russians experienced severe food shortages in 1921, 1941, 1961, and 1981.[25] Yet the party justifies the entire socioeconomic system as the wave of the future and as locked in mortal combat with capitalism, its dangerous but declining adversary. Consequently, despite an extremely rigid hierarchical structure, the leadership cannot feel secure and must continuously strive to demonstrate its legitimacy, often through accomplishments

24. Unexpectedly, Molotov was "rehabilitated" in 1984 at the age of ninety-four.

25. Seweryn Bialer believes that economic conditions facing the Soviet leadership in the 1980s will be so acute as to force them to modify their efforts at expansion abroad. See Bialer, "Andropov's Burden: Socialist Stagnation and Communist Encirclement," a paper presented at the Silver Jubilee Annual Conference of the International Institute for Strategic Studies, Ottawa, Canada, September 8–11, 1983. But economic crisis has in fact been commonplace throughout Soviet history; thus, should such a modification take place it is likely to be tactical and temporary.

abroad in the absence of accomplishments at home. Political stability is thus a perpetual concern for those who direct this rigid society.

It is because the Russians have failed throughout their history to establish self-assured and prosperous conditions at home that they are driven to equate stability and security with political domination abroad rather than with the attainment of social justice or economic well-being. Therefore, Solzhenitsyn, Pipes, Ulam, and Kennan are all partially correct in their analysis. But it is the totality of attributes, all pointing in the same direction, that drives Soviet policy.

In this context, nuclear weapons take on great importance for the Soviet leadership. Most significantly, the enormity of their destructiveness and the certainty that any adversary would suffer incalculable damage should it attack Soviet territory provide the Soviet government with a deterrent force that no Soviet or Russian leader has ever possessed before. Given American possession of nuclear weapons, a formidable nuclear arsenal is the surest guarantee that Soviet Russia, wracked by war and devastation throughout its history, will never again be attacked. This arsenal, therefore, is critical to the security to which the leadership aspires. Nuclear weapons, besides enhancing security, are a supreme symbol of technological mastery, a codification of Soviet superpower status. Sophisticated nuclear forces show to the Soviet people and to the world that the Russians are not inferior, even to the United States, in the most demanding field of technological prowess. Possession of nuclear weapons, in fact, is one of the only means by which the leadership can demonstrate Soviet technological parity with the United States. Nuclear weapons are thus vital to the quest for self-esteem in which both Soviet leaders and citizens are constantly engaged.

This demonstration of technological prowess also bears directly on the leadership's need to legitimize its authority and ideology. After all, Russia was the place, according to Marx, where the revolution was not supposed to occur. Demonstrating the good fortune of its occurrence is a perennial task. By now many Soviet citizens probably cannot imagine that any political system governing Russia other than the communist regime could have reached such military and technological heights. Where this is not the prevalent view, the government and the party try to see that it becomes so. In short, a powerful nuclear arsenal helps preserve the existing political order.

The role of the military in Soviet life also relates directly to the domestic significance of nuclear weapons. More than twenty years ago

Merle Fainsod, a pioneer in American studies of the Soviet Union, said: "The top command (of the army) is composed exclusively of Communists. Promotion and advancement depend on identification with the system. The Soviet officer corps occupies a privileged position in the Soviet social system; its material advantages and the honors accorded it make it one of the most attractive havens in Soviet life."[26] There is no evidence to suggest that conditions are any different today.

The responsibility for managing and operating the Soviet land-based missile program, the most potent arm of the nuclear arsenal, rests with the strategic rocket forces (SRF). The SRF was originally drawn from Soviet army artillery units, traditionally the most prestigious elements in the Russian armed forces.[27] The prestige of the SRF means that it can recruit the most talented personnel the system can produce, make large claims on the defense budget, and play a major role in formulating Soviet military strategy.

Moreover, the accumulation of large numbers of powerful weapons is consistent with the traditional Russian penchant for military conservatism. Hedging against surprise is an instinctive trait among armed forces, and, given Russian history, it is especially strong in the Soviet system. The adage "better safe than sorry" dictates the acquisition of forces beyond the level that non-Russians might think necessary for Soviet security.

Finally, nuclear weapons can potentially be used as instruments of coercion to further Soviet foreign policy objectives. From the end of World War II through the 1960s, the Soviet Union had to compete with the United States from a highly disadvantageous position. Faced initially with an American nuclear monopoly, the Russians remained significantly inferior to the United States in virtually all measures of nuclear weapon effectiveness until the late 1960s, although U.S. leaders often did not perceive this to be the case at the time. During this period, the challenge for Soviet strategists was to move forward without provoking the threat or the execution of an American nuclear attack. Using caution and bluff and carefully monitoring America's willingness to take risks, Soviet

26. Merle Fainsod, *How Russia Is Ruled* (Harvard University Press, 1963), p. 499.

27. Oleg Penkovskiy, the celebrated Soviet intelligence officer who spied for the West and was subsequently executed in 1963, observed in his papers: "Artillery has always been in a privileged position in Russia. From the time of Peter the Great we Russians have always been considered excellent artillerymen." See Penkovskiy, *The Penkovskiy Papers* (Doubleday, 1965), p. 33.

decisionmakers sought above all to avoid being the objects of American nuclear coercion. Under no conditions, however, must the leadership permit the United States to regain the position of nuclear superiority it held before the late 1960s.

The picture of the contemporary Soviet Union is surprisingly clear. Its political system is directed by an aged leadership on behalf of a privileged elite—the *Nomenklatura*—that numbers perhaps 50,000 people. Its economic system is wracked by chronic deficiencies. Its armed forces have experienced conflict since World War II only in neighboring communist states—to squelch riots in East Germany and Poland in 1953; to invade Hungary in 1956; to invade Czechoslovakia in 1968; to engage in border clashes with Chinese forces in 1969; and to invade Afghanistan in 1979. It has been unable to gain any political advantage from its sizable investment in nuclear forces since the waning of détente in the late 1970s; indeed, it is fearful of falling rapidly behind the United States in the military competition once again as this competition becomes increasingly high technology–intensive and makes greater and greater demands on scarce Soviet economic resources. It remains governed in its foreign policy by a willingness to probe but an aversion to risk. Incremental expansion of Soviet influence externally is seen by the leaders as a means of retaining their power internally. They do not seek, therefore, to shatter the nuclear stalemate and would probably be motivated to authorize nuclear weapon use only if they thought their existence depended on it.

The 1970s were a transitional period in which the Soviet Union reached nuclear parity and pursued détente with the United States and the West with greater vigor than in the past, primarily for economic reasons and to isolate China. As détente has cooled because of the American reaction to both the Soviet military buildup and Soviet policies in the developing world, the 1980s and beyond present new challenges. Now the task of the Soviet leadership is to see if its nuclear forces can help achieve goals such as the loosening of ties between the United States and Europe.

The pervasive Soviet insecurity and the ways nuclear weapons help alleviate it present the United States with a fundamental policy dilemma. If the United States does not compete vigorously in the nuclear arms race with the Soviet Union, it will encourage the aggressive tendencies of Soviet policymakers and jeopardize its own security and that of its allies. Moscow will see lack of American initiative and aggressiveness as signs of weakness to be exploited to Soviet advantage. If, however, the United States adopts a solely belligerent tone toward the Soviet Union, emphasizing American military power, Moscow will believe that its worst fears about the American threat are confirmed. Such a posture will be met by instinctive Soviet defensiveness and will stimulate Moscow to invest more heavily in nuclear and conventional force programs, regardless of the harmful effects on the civilian economy.[28] Moreover, a confrontational American posture would encourage the most repressive instincts of the Soviet leadership toward its own citizens.

To deal with such a policy dilemma, the United States needs to maintain a consistent, sophisticated, and balanced approach and shed the illusion that the Soviet problem can be solved. But as seen in the next chapter, the American national character is particularly ill suited to such a posture.

28. This characteristic of the U.S.-Soviet relationship should not be confused with the "security dilemma" in which potential adversaries are locked into situations in which "the policies of cooperation that will bring mutual rewards if others cooperate may bring disaster if they do not." See Robert Jervis, "Cooperation under the Security Dilemma," *World Politics*, January 1978, p. 167. From my analysis of the Soviet system I conclude that the only choices for the United States are to nullify Soviet power or to be dominated by it.

CHAPTER THREE

The Persistent Resurgence of the American Ideal

What kind of peace do I mean? What kind of peace do we seek? Not a Pax Americana enforced on the world by American weapons of war. Not the peace of the grave or the security of the slave. I am talking about genuine peace, the kind of peace that makes life on earth worth living, the kind that enables men and nations to grow and to hope and to build a better life for their children—not merely peace for Americans but peace for all men and women—not merely peace in our time but peace for all time. John F. Kennedy

THE DOMINANT American trait influencing public policy is a dual personality that simultaneously seeks a military posture second to none and a denuclearized world permanently at peace. Because of this dual personality, the United States has adopted radically different and often inconsistent policies with unrealistic premises and cannot muster enough domestic political support to sustain them.

To understand the American character, it is necessary to understand its penchant for "concepts" as guides to policy. These concepts have been more than slogans; they have been expressions of commitment, often consistent with capabilities and presumably always consistent with intentions. For a conception of national security policy to be credible both in a democratic society and to that society's allies and potential adversaries, it must be consistent with the material strengths of the nation and the collective will of its people. A conception inconsistent with material strength would weaken alliances and be ignored by adversaries. A conception markedly at variance with the will of the people would ultimately be overturned by public ridicule and opposition.

There are, of course, tensions between the general and the specific in the conceptualization of policy. Too much abstraction leads to meaningless phrases that cannot be tested against reality; too much specificity sacrifices comprehensiveness and fails to take account of the unpredictable aspects of human affairs. Although no general formula for American security policy can be derived that is analytically as rigorous as, for example, the Newtonian relationship between mass and acceleration, the total absence of conceptualization—the sheer promotion of policies based on "ad hocracy" and expediency, resistant to any form of generalization—invites confusion, inconsistency, and miscalculation by friends and foes alike. The history of the American experience tells us that what must be found in a formulation of policy are not only concepts but also guidelines for action, criteria that, once met, suggest unambiguous policy initiatives.

The American Character

The cultural context of American politics has had a great deal to do with the shaping of policies. Prenuclear America was a stark contrast to Russia. The United States, from the days of the founding fathers, has been intentionally fragmented so that no one segment or branch of government could dominate policy without the acquiescence of the others. This separation of powers was not pious rhetoric but a deliberate effort to safeguard democracy—even at the expense of efficiency and consistency—and to protect the strong states from dominance by the federal government. Religious freedoms were generally well protected. Also, while the separation of church from state was for the most part scrupulously respected, moral principles derived from the Judeo-Christian ethic found their way into American policy.

If today the words of Jefferson, Lincoln, Woodrow Wilson, and Franklin D. Roosevelt sometimes seem trite or moralistic, it is because they reveal the dual characteristics of American political life: a striving for national economic, military, and diplomatic self-interest on the one hand and universal principles of political freedom and human rights on the other. Americans have never quite figured out where pragmatic concerns leave off and universalist ideals take over and have debated the point in the policy arena unrelentingly for two centuries.

By the end of World War II, Americans were infused with an enormous sense of national pride and superiority, a result of U.S. achievements both at home and abroad. Contrary to the Russian experience, American territorial expansion had produced secure and definite borders—a weak Mexico, a seemingly assimilated Canada, and two oceans. In the century between the Mexican War and the end of World War II, the United States had transformed itself into the world's foremost industrial, agricultural, and military power. The country did not experience any foreign setbacks. Political change at the national level, despite the assassinations of Lincoln, James Garfield, and William McKinley, through its very regularity affirmed that the democratic process worked. Nowhere on earth was openness in debate more vigorous and absence of coercion by the state more complete.

The following observation of Tocqueville's, therefore, was insightful but incomplete:

> The Anglo-American relies upon personal interest to accomplish his ends, and gives free scope to the unguided strength and common sense of the people; the Russian centres all the authority of society in a single arm: the principal instrument of the former is freedom, of the latter, servitude.[1]

He perceived the strength of the American system and the limitations imposed by the czars, but he failed to articulate several characteristics of American culture that do not serve the United States well in contemporary international affairs. These characteristics are an absence of historical perspective, a problem-solving orientation, and a preference for the tactical rather than the strategic.

The absence of historical perspective is a consequence of long periods of isolation from much of world and from European politics. This lack of perspective causes American policymakers to often jump into the middle of policy differences between countries without adequate appreciation of the deep and complex origins of their differences. At best Americans are guided by a few selected historical landmarks, such as treaties, crises, and assassinations, as the basis for formulating U.S. policy. The orientation toward problem solving steers Americans toward solving international disputes rather than improving and effectively managing permanent conditions. Because of Americans' penchant for completing a job so they can move on to the next, they tend to view international conflicts as having beginnings, middles, and endings. Once they think an end has been reached, they have little patience for giving continuing attention to the situation. Emphasis on the discrete rather than the continuous is the essence of the American approach to international relations. As a result of these first two characteristics, Americans tend to concentrate on the pieces of a particular foreign policy puzzle rather than on either the interrelationships of the pieces or the need to arrange or integrate them into a preferred, coherent whole.

1. Alexis de Tocqueville, *Democracy in America* (Washington Square Press, 1964), p. 125.

Although the term *strategy* is constantly invoked to describe particular American actions or doctrines, it is *tactics* that most of them are really about. Strategy takes political and psychological as well as the more easily quantifiable economic and military considerations into account. Coordinated initiatives or actions in the service of overarching objectives are the true hallmark of strategic behavior. Yet uncoordinated *re*actions, with little or no advance planning, largely in response to unanticipated developments, dominate the actual record of American foreign policy.[2] President Reagan's insertion and then abrupt withdrawal of U.S. marines in Lebanon in 1984 is a fine example of a tactical initiative reflecting an ahistorical perspective and a problem-solving orientation.

These characteristics were particularly ill suited for the formulation of policy after World War II. Since 1945 the United States has been thrust, and has sought to be thrust, into a leadership role in world politics, pitted against an adversary with both a long and deep historical memory and an elaborate, albeit imperfect, sense of the strategic. American solutions for temporary problems would not easily serve the fundamental challenge of the nuclear age: the effective management of a permanent condition that involves nuclear deterrence and political competition between adversaries.

Containment, Vietnam, and Its Aftermath

With these limitations inherent in the American character, the United States proceeded to deal with the postwar world by formulating and implementing the policy of containment, which George Kennan described as "the adroit and vigilant application of counterforce at a series of constantly shifting geographical and political points, corresponding to the shifts and manoeuvres of Soviet policy."[3] The policy of containment dominated American foreign policy from the late 1940s to the mid-1960s, because it was implemented in a reasonably consistent fashion and because it worked. Whatever the intentions of its authors—and

2. This last point differs sharply from the judgment that "American foreign policy has so often been virtually absorbed in strategy." See Stanley Hoffmann, *Gulliver's Troubles, or the Setting of American Foreign Policy* (McGraw-Hill, 1968), p. 148. This point excepted, Hoffmann's complex dissection of American behavior is compelling. See, in particular, "The National Style: An Analysis," in *Gulliver's Troubles*, pp. 94–175.

3. "X," "Sources of Soviet Conduct," *Foreign Affairs*, vol. 25 (July 1947), p. 576.

there remains considerable dispute about what these intentions were—containment came to mean a network of formal military alliances and bases and a tendency to use military force to prevent the expansion of communist influence beyond those territories seized by the Red Army at the end of World War II.[4]

Containment defined in these terms, except in the notable cases of China and Cuba, was a success; despite acrimony during the Korean War, it commanded substantial support of all important segments of the American body politic—businessmen, politicians, journalists, labor leaders, scholars, and the military. It was seen as global rather than region specific in its applicability and provided the fundamental rationale for American political and military initiatives in international affairs.

America's traumatic experience in Vietnam, however, shattered public confidence in containment, at least as it had been applied well into the 1960s. To be sure, the lessons of Vietnam are different for different constituencies, as lessons from history always are. But there is little doubt that the "no more Vietnams" sentiment—that is, the belief that the United States should fight to win quickly or stay out—is now a principal reference point for American foreign policy, especially in relation to the developing world. The influence of this sentiment is unlikely to disappear from America's political psychology unless and until a major new trauma is experienced or a staggeringly impressive achievement is demonstrated. For millions of Americans—including many who grew up during the Vietnam War and who will assume high positions of public responsibility at the end of this century—Vietnam

4. George Kennan, the principal author of the concept while in government service, subsequently criticized its application on the grounds that the military dimension was overemphasized at the expense of the political. But in his attempt to clarify his original intentions, he addressed the tension between concept generality and specificity in a confusing way. "What I said in the X-Article was not intended as a doctrine. I am afraid that when I think about foreign policy, I do not think in terms of doctrines. I think in terms of principles." George F. Kennan, *Memoirs (1925–1950)* (Bantam Books, 1969), p. 383. According to Webster, however, a doctrine is "a principle of law established through past decisions" or "a statement of fundamental government policy especially in international relations"; and a "principle" is "a comprehensive and fundamental law, doctrine, or assumption." Presumably, what Kennan meant was that he had set out certain guidelines for government action at a rather high level of generality that in practice were applied selectively and incorrectly.

It is not the intention here to review the twists and turns of the intellectual and policy debates surrounding the policy. This is admirably assessed in John Lewis Gaddis, *Strategies of Containment: A Critical Appraisal of Postwar American National Security Policy* (Oxford University Press, 1982).

now rivals or supplants Munich and Pearl Harbor as the dominant memory, the most relevant analogy, to guide the future course of U.S. foreign policy.[5]

Even in the face of growing conservatism in the United States since the mid-1970s, consequences of the Vietnam War remain vividly evident. Many Americans, mindful of the excessive violence of the South Vietnamese regime against its own people, no longer believe that support for authoritarian regimes against their communist challengers should be an automatic U.S. policy preference. The intensity of the conflict within the communist world, principally between China and the Soviet Union but also between China and Vietnam, has weakened the American image of a monolithic communist threat. This revised image, along with a new appreciation of the ruthlessness and corruption of many noncommunist governments in developing countries, has made many Americans less confident about which side to choose, if any, in third world conflicts.

The asymmetry of stakes has also emerged as an important lesson from Vietnam. In Southeast Asia the United States was engaged in a limited war and used limited means to achieve limited ends. The local adversary, however, was waging unlimited war, using all the means at its disposal to achieve nothing less than total victory. This asymmetry of stakes was reflected in an imbalance of incentives and willingness to incur costs. Over the long haul the military balance of power was less important than the disparity in political motivation. Moreover, the power of television to bring into American living rooms the horror of war, and to influence attitudes about the war, was fully demonstrated and is a

5. This is an argument, admittedly contentious, that the post-Vietnam period of U.S. foreign policy will continue for many years, perhaps decades, beyond American disengagement from Southeast Asia, and that assertions of the arrival of a post-post-Vietnam phase in U.S. policy fail to appreciate the very deep psychopolitical scars this experience has left, not only on Americans, but on peoples everywhere. Since the communist victory in Vietnam, for example, the "Vietnam analogy" has been consistently invoked by skeptics of American military involvement in the Middle East and Central America. An important work on the patterns and dangers of policy reasoning by historical analogy is Ernest R. May, *"Lessons" of the Past: The Use and Misuse of History in American Foreign Policy* (Oxford University Press, 1973). A potentially important application of the "lessons of Vietnam" for future American policy is contained in a 1984 speech, "The Uses of Military Power," delivered by Secretary of Defense Caspar W. Weinberger before the National Press Club. See "Excerpts from Address of Weinberger," *New York Times,* November 29, 1984. Weinberger stated that "once a decision to employ some degree of force has been made, and the purpose clarified, our Government must have the clear mandate to carry out that decision until the purpose has been achieved."

salient factor to be weighed carefully by U.S. policymakers contemplating military intervention.

The doctrine that a protracted, limited war could be waged effectively through controlled escalation—at least as practiced by a democratic state that broadcasts its self-criticism through the news media—was also discredited. The lesson was: when blood as well as treasure is involved, open-ended commitments cannot be sustained if the objectives are not compelling.

The importance of regional differences is still another lesson of the Vietnam experience. Containment can be criticized not only for its excessive reliance on military force, but also for its assumption that it can be applied anywhere, irrespective of the political, economic, cultural, historical, or topographical characteristics of a particular region. It is now known that these characteristics do make a difference and that American military power and technological prowess cannot be applied in markedly different settings with the same probability of success.

The Vietnam debacle, moreover, not only inhibited the American tendency to intervene militarily in distant conflicts, but also dampened the enthusiasm of third world governments for overt American support. The death and destruction associated with American military intervention, the questionable reliability of U.S. security commitments, the decline in American willingness to provide security and economic assistance, and the liability of being associated with a power that is seen as supportive of the status quo and resistant to progressive economic and social development were all serious warning signals for third world leaders. In Asia, Africa, and Latin America, even where opposition to the Soviet Union is pronounced, most governments came to believe that to have America as a physically and diplomatically visible ally is a luxury few of them can afford.[6]

The foreign policy of the United States went into a tailspin after Vietnam, and the decline in American self-confidence that followed the war led to a polarization of views about the future direction of U.S. policies abroad. Decline in American self-confidence had begun with domestic events: the assassinations in the 1960s of President John F. Kennedy, Martin Luther King, and Senator Robert F. Kennedy. Amer-

6. It will be interesting in this context to evaluate the long-term effectiveness of the Reagan administration's military commitments in the Middle East, Central America, and the Caribbean, which are clearly designed to shake America loose from the constraints imposed on itself by the Vietnam memory.

icans were especially traumatized by the murder of President Kennedy. In 1963 relatively few Americans could recall the last presidential assassination—the shooting of President William McKinley at the Pan-American Exposition in Buffalo in 1901. What happened sixty-two years later on that fateful day in Dallas and the lack of a satisfactory explanation for why it happened have left a scar on America's political psychology that will last at least as long as the lives of those who witnessed the assassination's four-day televised aftermath. Kennedy, despite the questionable merits of his administration's record and the facts about his personal conduct that surfaced after his death, was a leader who, through rhetoric, public posture, and selected policy initiatives, generated enormous self-confidence among the American people. His youth, energy, articulateness, and good humor captured the imagination of countless Americans and non-Americans alike. His sudden and tragic death shattered not only what we thought *he* could become but also what we thought *we* could become. Inspiration is the essence of leadership; in this respect, the United States was virtually leaderless from Kennedy's death to the ascendancy of Ronald Reagan as president.

With the loss of Kennedy and the weakening of support for worldwide containment, policymakers offered alternative approaches to guide U.S. actions abroad. Some argued that the underlying concept of American foreign policy should be balance of power. To achieve this the United States would adhere to détente with the Soviet Union, seek to establish in developing countries rules of conduct accepted by both superpowers, and maintain regional stability by relying on selected regional hegemonies. But this "structure of peace" designed by President Richard Nixon and his assistant for national security, Henry Kissinger, wobbled and then collapsed in the face of the unrelenting growth of Soviet military power, the use of Soviet proxy forces to promote regional political change, the inherent political instability of several key regional hegemonies with which the United States was allied, the passage of the Jackson-Vanik Amendment linking Soviet most-favored-nation status to Jewish emigration, and the inability of the concept of balance of power—because of its intrinsic ambiguities—to command and sustain consensus among the American political elite. Ironically, Nixon and Kissinger's efforts to turn back the sails of containment to compensate for reduced American capabilities were halted before the United States had an opportunity to determine their effectiveness.

Others sought a formula in no formula, urging instead an ad hoc approach that expressly eschewed any form of conceptualization, on the grounds that any foreign policy concept would fail to capture the complexities of the modern world and would lead to disaster, as containment had led to the Vietnam War. It was thought desirable to determine instead what our foreign policy objectives were and resolve problems as they arose, issue by issue. But this approach, as practiced by the administration of President Jimmy Carter, understandably led to marked inconsistencies and incoherence that surprised adversaries, angered allies, promoted cynicism among nonaligned states, and eventually provoked confusion and disillusionment at home.

Most recently, others have sought to reconstitute containment, either in its original formulation or in a more limited anti-Soviet guise, as the guiding principle of American foreign policy. This approach, however, not only ignores the very real consequences of our experience in Vietnam but also fails to take into account the significant decline in U.S. economic and military power that has taken place in the last two decades.

American Domestic Constraints

Although it would be a gross distortion to describe the years from 1947 to 1966 as the golden age of U.S. foreign and defense policy, during that period the nation did possess four crucial assets: an organizing concept to guide its behavior in the world, military superiority over its principal adversaries (though not on the ground in central Europe), extraordinary economic strength, and a considerable degree of national self-confidence. These strengths were reflected at home in a national consensus on most major foreign policy initiatives, in an executive-congressional relationship on foreign and defense policy that bordered on the hierarchical, and in a pervasive confidence in American society and American technology that crowded out the most parochial policy motivations and preferences.

By the early 1970s three significant realizations began to dawn on the American people. The first was that the United States and the Soviet Union were not engaged in a global contest for power that was "zero sum" in nature. Americans began to believe, influenced in part by Soviet behavior and by the emerging Sino-Soviet split, that limited cooperation with the Soviets was possible and surely preferable to unbridled competition and that the most important shared value was the desire to avoid nuclear war.

A second widely adopted view, which stemmed from the early writings of certain American defense experts, was that numbers of nuclear weapons no longer mattered, or mattered very little. It was felt that the conditions for nuclear deterrence would be satisfied once U.S. leaders believed that a few thousand warheads could survive even the most well-orchestrated first strike and be delivered in a resounding retaliatory attack. The buildup of forces beyond this point seemed to serve no discernible military or political purpose.

The third emerging judgment was that military force was of limited value in achieving foreign policy goals, a lesson learned painfully in Vietnam. It became fashionable to believe that economic and other nonmilitary concerns would come to dominate a world of growing interdependence and that traditional definitions of security were becoming passé.

The prevalence of these views paved the way for domestic support of negotiated arms control with the Soviet Union and permitted Nixon and Kissinger to use arms control as part of their strategy to enmesh the Soviet Union in the complexities of the international community. There the Soviets would find that, to achieve their foreign policy objectives, they would have to make trade-offs. In this way American leaders hoped to complicate Soviet decisionmaking and, ultimately, to reform Soviet behavior and reduce Soviet aggressiveness and expansionism.

The dispute that arose over the numerical disparities between Soviet and American land-based and sea-based missile launchers contained in the SALT I Interim Agreement signaled that numbers (of warheads or launchers) do matter, at least in the context of the American defense debate. It is fine for Americans to believe in parity as long as they enjoy superiority. But when Americans see numerical parity as having been achieved and then rapidly slipping away, they begin to question parity's validity. This is what happened in the 1970s, as the U.S. defense debate became more intense at home and abroad: the numbers became at least politically significant; American leaders became less confident; our allies became more uncertain about the credibility of U.S. support; and the Soviet Union grew bolder in its attitudes, pronouncements, and policies. In the 1970s, despite SALT, the strategic nuclear balance as measured by most static indicators shifted dramatically to Soviet advantage, and the assumption that numbers do not matter was politically discredited.

After the SALT I agreements were ratified, Soviet nuclear force deployments and Soviet policy in developing countries undermined the assumption that the Russians and the Americans could cooperate. The litany of Soviet "transgressions" is by now familiar: the deployment of high-accuracy intercontinental ballistic missiles (ICBMs), which pose a serious threat to the U.S. ICBM force and its command, control, and communication system; the deployment of high-accuracy intermediate-range ballistic missiles (IRBMs) for use in the European theater, a deployment that could nullify NATO's ability to control escalation in the event of a conventional war in Europe; the testing of antisatellite weapons; questionable adherence to certain provisions of the ABM (antiballistic missile) Treaty and other arms control agreements; the development and maintenance of an elaborate civil defense program; the plentiful military writings that focus on nuclear war fighting; the use of Cuban proxy forces and Czech and East German military advisers in sub-Saharan Africa; the invasion of Afghanistan; and the orchestration of political repression in Poland. People may disagree about the significance of any one of these developments, but their cumulative effect has been to paint a picture for many Americans of an adversary bent on destroying the West, an adversary that shares nothing with the United States, perhaps not even the wish to avoid nuclear war.

Moreover, after the United States disengaged itself from Vietnam, crises and conflicts continued to materialize, which accentuated American weakness. The avoidance of a Soviet-American confrontation during the 1973 Arab-Israeli war; American inability or unwillingness to counter communist gains in Angola, Mozambique, and Ethiopia; the humiliating hostage situation in Iran; and the smoldering problems of political instability in Saudi Arabia, South Korea, and the Philippines all underscored the significance of the role of force in the modern world. This renewed appreciation for military might produced a marked shift in congressional and popular attitudes about the need for a larger defense budget, rapid deployment forces, a more potent naval presence to project American power, and even the resumption of military conscription.

But this attitude shift in favor of heightened military force could not be readily translated into coherent policies and programs because of tensions between the U.S. Congress and the executive branch. These tensions were reflected in Congress's rejection of presidential leadership through the War Powers Act, as a consequence of Watergate and other perceived excesses of the "imperial presidency." In addition, severe limitations on congressional leadership resulted from the breakdown of the seniority system and the large influx of newly elected officials who preferred political individualism to party loyalty.[7]

Another barrier in the way of formulating and implementing sensible, coordinated foreign and defense policies is the sheer overworked condition of the senior figures in the national security policy community. The overloading of responsibilities on a few key officials, it has been observed, has produced several harmful effects:

> among them crisis management as a way of life and reaction to events rather than planning and initiatives; inadequate staffing of positions simply because the principals usually lack the time either to determine what kind of support they need or to absorb the results of the work that is done; the ability to handle only a very small number of issues more or less simultaneously; the tendency of the staffs to coordinate and compromise at the lowest common denominator rather than let issues escalate to their harried principals; the exploitation of this process by the ambitious to make their mark, expand staffs, and attract notice; and leaks as disaffected staffs become alienated from this interminable process or are unable to make their views heard.[8]

This problem is just as severe at the highest levels in the Department of State and in the office of the assistant to the president for national security affairs. With their backbreaking schedules, it is little wonder that government's top officials produce contradictory and unsatisfactory policies.

More obstacles abound. One is the fundamental differences in outlook, amounting to essentially irreconcilable ideologies, held by key players who influence national security policy. This obstacle was especially pronounced under President Carter. In the late 1970s, officials with impressive credentials trained their talents on foreign policy, defense, and arms control issues. These officials were graduates of America's leading universities, and they collectively represented expertise on a wide range of legal, political, military, and technical issues. But they disagreed sharply on virtually every fundamental question about national security policy: the delicacy of the strategic nuclear balance; the political and military significance of growing U.S. ICBM vulnerability; the political utility of nuclear weapons; whether the Soviet Union was motivated primarily by defensive or expansionist aims; whether

7. See James L. Sundquist, "The Crisis of Competence in Government," in Joseph A. Pechman, ed., *Setting National Priorities: Agenda for the 1980s* (Brookings Institution, 1980), pp. 531–63.

8. William W. Kaufmann, "Defense Policy," in ibid., p. 285.

Soviet foreign policy was opportunistic or based on a grand design; whether Soviet leaders were optimistic or pessimistic about the future and whether, consequently, a "window of vulnerability" would materialize in the 1980s; the meaning of strategic stability and the desirability of deploying high-accuracy counterforce weapon systems; what SALT had accomplished; and what negotiated arms control with the Soviets could accomplish in political, military, and economic terms.

The deeply held views on these issues gave rise to perceptions and self-perceptions that accentuated polarization. Some experts saw themselves as latter-day Churchills doing battle against dangerously naive Chamberlains; others saw themselves as sophisticated foreign policy centrists at odds with simplistic, Neanderthal-like extremists who saw only military solutions to what were most often political problems. Consequently, the policy process was marked by intense bargaining, most often resulting in outcomes that reflected trade-offs and contradictions rather than consistency of purpose.

The conflicting organizational stakes represented in the policy process dramatized differences in perspective. The different priorities of the policymaking bodies—the Departments of State and Defense, the assistant to the president for national security affairs and the staff of the National Security Council, the Arms Control and Disarmament Agency, the Joint Chiefs of Staff, the Office of Management and Budget, the Office of Science and Technology Policy, and the president and his staff—made consistent policymaking and implementation that much more difficult.

Another major constraint on the policy process is what might be called the clean-slate phenomenon. Every new administration tries to make its mark by evaluating the major policies and programs it has inherited, by dismantling ineffective programs, and by introducing as many distinctively new initiatives as the system will tolerate. Wiping the slate clean and starting fresh guarantees a certain degree of discontinuity, real or perceived, from administration to administration. Given all the above constraints, it would be naive to expect the current policy process to produce consistent foreign policy initiatives and defense programs on a sustained basis.

By 1980 not only had the United States seemingly lost track of the organizing concept that had structured and rationalized American foreign policy, but it also saw itself as slipping behind the Soviet Union militarily in the face of relentless, though not necessarily effective, Soviet expansionism. Americans and others began to perceive the United States as an economic power in serious decline. And American self-confidence began to wane. The victorious and glorious image Americans had of their country after World War II was replaced by that of U.S. embassy officials fleeing from the North Vietnamese in the last helicopter to leave Saigon and by the wreckage of American helicopters in the Iranian desert, a symbol of the futile and inept attempt to rescue American hostages taken at the U.S. embassy in Tehran.

By 1980 the United States had become a wounded nation trying to find answers and relief in simplicity, panaceas, and jingoism. Presidential authority eroded. Specialized interests dominated congressional debates on weapon-procurement decisions. Confusion reigned on the central precepts of American military strategy. And voices in the public debate deprecated U.S. capabilities and elevated Soviet effectiveness far more than the facts warranted.

Negativism has a momentum of its own that can be broken only by success. Fresh leadership and competence, a major technological or conceptual breakthrough, or even a successfully handled international crisis would go a long way toward restoring American national confidence. Without such psychological renewal, however, no investment in material strengths can, by itself, give coherence and purpose to U.S. defense policy.

Ronald Reagan has succeeded in rectifying many, but not all, of these deficiencies. His foremost accomplishment in his first term as president was to restore American self-confidence.[9] He achieved this by reviving the national economy (despite increasing unprecedented budget deficits); enhancing the perceived strength of the nation's defenses by concentrating on weapon system procurement; winning a quick victory in Grenada; talking sometimes tough and sometimes reasonably to the Russians; raising the prospect of regaining military superiority over the Soviet Union while simultaneously endorsing equitable arms control agreements; working effectively with Congress on selected issues he sought to influence; and possessing a great talent for communication and an appealing personality. He was assisted in his task by two fortuitous developments: first, the unprecedented situation in which three sick and elderly Soviet leaders—Brezhnev, who died in 1982; Andropov, who died in 1984; and Chernenko—presided over a Politburo dominated by

9. See "America's Upbeat Mood," *Time*, September 24, 1984, pp. 10–17.

caution and consensus; and second, a global oil surplus that weakened energy prices and facilitated efforts to reduce inflation. Reagan was not able, however, to manage effectively the deep divisions within his administration over arms control policy, an obstacle that may come back to haunt him in his second term.[10]

The American Character and the Role of Nuclear Weapons

Americans are deeply divided in their judgments about the appropriate level of military strength for their country and how much nuclear weapons should contribute to that strength. Moreover, this split is very evident in official policy, influences Congress, and is occasionally embodied in an individual president. As a result the United States has been unable to develop, implement, and sustain a consistent nuclear policy.

Americans who consider the two decades after World War II as the norm are disturbed about the later slippage in American power. They see the United States as possessing the technological ingenuity and economic resources to be the most powerful nation on earth and believe that modern sophisticated nuclear forces are essential ingredients in achieving this largely political objective. Because nuclear weapons and their delivery vehicles are at the frontier of military technology, they symbolize the American affinity for and self-confidence in technological innovation. Many Americans and much of the American leadership derive considerable satisfaction from the fact that their armed forces possess the most advanced and most lethal weapon systems of any nation.

It is also widely recognized that nuclear weapons play a decisive role in deterring war with the Soviet Union, in particular a nuclear first strike by Soviet forces on the continental United States. A judgment that commands broad support is that the threatened use of these weapons in retaliation is an imperfect but effective way to ensure that they will not actually be used.

The political and psychological usefulness of nuclear weapons in

10. The definitive account of the intense bureaucratic rivalries within the administration concerning nuclear arms control policy toward the Soviet Union is Strobe Talbott, *Deadly Gambits: The Reagan Administration and the Stalemate in Nuclear Arms Control* (Knopf, 1984).

crisis situations is also often cited as evidence of their importance. A lively and unresolvable debate rages to this day about whether local naval superiority, a greater nuclear capability, or a combination of these U.S. advantages produced an outcome in the 1962 Cuban missile crisis that virtually all Americans consider a triumph for the United States.[11] Few would wish that at that time the United States had swapped its nuclear arsenal for that of the Soviet Union. The lesson seems to be that, although having a decisive edge in nuclear weapons may not influence some crises and may positively influence others, it can never hurt the holder of the superior force. Consequently, many consider the potential coercive power of nuclear weapons in nonwar situations to be nonnegligible.

Organizational incentives also reinforce support for U.S. nuclear weapon programs. In the military, nuclear weapons and their use occupy center stage. One of the main purposes of the air force is to be able to deliver thousands of nuclear weapons on an enemy's homeland using both ICBMs and intercontinental-range bombers. The crews that serve on submarines carrying ballistic missiles and on board dual-capable carrier-based aircraft (that is, those able to deliver nuclear and nonnuclear weapons) are among the most elite and prestigious in the navy. In the army, use of short-range and intermediate-range nuclear systems has been, at least doctrinally, central to the maintenance of security in Western Europe. These systems give decisionmakers the option to respond flexibly to an attack by Warsaw Pact forces depending on the circumstances of the attack.

Nuclear systems are as central to the technical community as they are to the military. Each year the nuclear weapon laboratories in Los Alamos, New Mexico, and Livermore, California, recruit some of the finest young scientists and engineers that American higher education can produce, and many other highly qualified people obtain work in private industry, where they assist in the research, development, and production of nuclear weapons and their delivery vehicles. Indeed, the essence of American teamwork and efficiency is symbolized more by the Manhattan Project that produced the first atomic bomb than by any other single technological achievement in this country's history, with the possible exception of the first manned landing on the moon.

11. Several of the key American participants in the Cuban missile crisis now claim that U.S. nuclear superiority was not a significant consideration. See "The Lessons of the Cuban Missile Crisis," *Time*, September 27, 1982, pp. 85–86.

There is another side to the American attitude, one of guilt, fear, and moral concerns. Many distinguished scientists, for example, feel guilty about the dropping of the first atomic bomb on Hiroshima and the subsequent devastation of Nagasaki. They feel these were indefensible acts that should never have been committed and can never be justified. To many Americans the accumulation by both superpowers of thousands of nuclear weapons, many with yields one or two orders of magnitude greater than the bomb dropped on Hiroshima, seems truly terrifying and totally divorced from any rational political or military purpose. Moreover, as noted in the Catholic bishops' pastoral letter on war and peace, there are profoundly serious moral reservations associated with even the threat to use nuclear weapons, much less their actual use.[12] Because of these concerns, political support has developed for a bilateral and verifiable freeze on the development and deployment of U.S. and Soviet nuclear weapons and their delivery vehicles.

The tensions, therefore, within the American body politic on nuclear weapon issues are profound. Some see nuclear weapons as protecting the United States from the Soviet threat, while others see their possession as the principal threat. Some see the use of nuclear weapons to gain political advantage over the Soviet Union as in the national interest, while others see this course as leading to mutual suicide. Some see nuclear weapons as serving a multiplicity of foreign policy objectives; other see them as serving nothing at all. Some see nuclear weapons as a great technological achievement; others see them as a technological monster. Some want modernized nuclear forces as a hedge against uncertainty, while others say, "Enough is enough."

The effect of these differences in policy terms is not that Americans have no consensus but rather that they move from one consensus to another. The lack of broad public support for American military programs in 1976, the pronounced support for them in 1981, and the waning of this support in 1982 illustrate the continual adoption and subsequent rejection of temporary palliatives for problems that do not go away.

An even more significant phenomenon is now occurring, however. The Vietnam War and the deployment of Soviet intercontinental nuclear forces, along with the decline of American economic primacy, have shattered the American sense of invulnerability that had long been

12. See "The Challenge of Peace: God's Promise and Our Response," *Origins: NC Documentary Service,* vol. 13 (May 19, 1983) (publication of the National Catholic News Service).

central to national self-confidence. The search to regain this invulnerability haunts many Americans and is at the core of the national debate. For some, this ideal requires not only protecting American self-interest but also promoting a harmonious world order based largely on the principles of American democratic liberalism. For others it means confronting the Soviet threat and somehow rendering it impotent.

American public support for both strength and peace translates into a long-term commitment to both nuclear force modernization and nuclear arms control. Presidential positions that deviate too far from this contradictory flow of American opinion will not endure for long.[13]

Reagan's approach to the Soviet Union and nuclear arms control can be seen as fully consistent with the cyclical American pattern—first confrontational, then accommodative—that has characterized U.S. national security policy since World War II. Reagan symbolized still another return to the American ideal, espoused by Wilson and embraced by many of his successors, of "making the world safe for democracy." Although his rhetoric toward the Soviet Union during his first two and one-half years in office was among the harshest uttered by an American president in the postwar period—and this greatly exacerbated relations with the Soviets—there were few deeds commensurate with his words. In short, public concern that the nuclear stalemate would be shattered by American recklessness or provocativeness proved to be groundless. The international situation was and is far more stable than it at times appeared to be.

13. A thorough examination of American public attitudes on nuclear weapons conducted during the 1984 presidential campaign documented this dual commitment: first, that the United States must not adopt any policy that the majority of Americans will perceive as "losing the arms race"; and second, that Americans are convinced that it is time for negotiations, not confrontations, with the Soviets. See Daniel Yankelovich and John Doble, "The Public Mood: Nuclear Weapons and the U.S.S.R.," *Foreign Affairs*, vol. 63 (Fall 1984), pp. 43–46.

CHAPTER FOUR

Increased Accuracy and Heightened Vulnerability

Though this be madness, yet there is method in't.
Shakespeare, *Hamlet*

FORTY YEARS after the first use of the atomic bomb the United States and the Soviet Union combined possess roughly 60,000 nuclear weapons and have deployed them on a variety of delivery vehicles. Yet no nuclear weapon has been used in warfare since 1945 and, except possibly in the Cuban missile crisis, superpower leaders have not come close to authorizing their use. This is an extraordinary record in light of the intense political rivalry between the two superpowers. This chapter examines the proposition that in the decades ahead, the deployment of high-accuracy weapons—an inevitable consequence of advances in information processing and related technologies—will necessarily heighten the perceived vulnerability of both U.S. and Soviet nuclear forces. This situation will in turn cause both countries to intensify their efforts to change the basic strategic framework that has kept the nuclear peace until now.

In evaluating the significance of emerging technologies it is important to keep in mind the interconnectedness of technology and policy. A state's basic objectives and the policies required to achieve them cannot, or at least should not, be divorced from the weapons available to those in command and on the battlefield. Indeed, weapon technologies can themselves alter the objectives of national leaders if they provide new and significant capabilities not previously thought attainable. In the evolution of military technology, several patterns place the nuclear arms race in historical perspective.

First, the history of warfare reflects a persistent competition between offensive and defensive weaponry. Measure, countermeasure, and then counter-countermeasure make up the recurring pattern. When stone replaced wood as the material for castle fortifications in fourteenth-century Europe, for example, castles became impregnable, and famine and disease were the only countermeasures against this defense-dominant environment. Similarly, the mounted, armored knight was a powerful offensive force until the advent, in the fifteenth century, of the long bow, the first armor-piercing weapon. Although this pattern of measure and countermeasure has continued unabated, the categories of offensive and defensive weaponry have often been misleading. The stone castle, a quintessential defensive weapon, prompted knights to make offensive forays into the countryside confident that they could then retreat to their impregnable fortresses. In other words, a powerful defensive capability has often encouraged offensive action.

Second, man has developed several "ultimate" weapons only to find that some new technology has nullified their effectiveness. What is noteworthy about the twentieth century is the rapid pace of technological innovation. Whereas before World War I it often took centuries for new technologies or materials to add to weapon development—the cannon, for example, was used for over three centuries—the last fifty years of technological innovation have greatly accelerated the obsolescence of modern weaponry.

Third, many revolutionary developments in weaponry were initially underrated and resisted before their true military significance was widely appreciated. From the time of the Italian city-states, when Niccolo Machiavelli ridiculed the handgun as being too inaccurate and unreliable for warfare, to the 1920s, when General Billy Mitchell had trouble selling the airplane to the U.S. Army, the revolutionary potential of new weapon systems has often met with the skepticism that results from routine procedure and modes of thought. Those able to overcome this skepticism the most quickly were the ones who benefited most from innovation.

Fourth, both simple advances in weaponry and nonmilitary innovations have sometimes revolutionized warfare. In the former category a striking example is the tank, which has been graphically described as "an invention relatively simple in conception, relying upon no new scientific ideas and no radically new technology, but simply upon the proper assembling of technical devices already long in use—this was the weapon that finally revolutionized land warfare and brought an end to the stalemate [of World War I]."[1] In the latter category are universal

1. Bernard Brodie and Fawn M. Brodie, *From Crossbow to H-Bomb* (Indiana University Press, 1973), p. 199.

military conscription—the draft—instituted by the French at the end of the eighteenth century and, in the nineteenth century, the railroad, which revolutionized military logistics and concepts of maneuver warfare.

Fifth, and most important, the experience of war from ancient times through the end of World War II reveals one overriding fact: most shots missed their targets. Weapons have always been categorized by their range, lethality, and accuracy. The history of military technology is the story of man seeking to perfect weapons of longer range and greater power and accuracy. Until the advent of nuclear weapons, man made the most progress in increasing weapon range and the least in improving accuracy. Accuracy was mostly an inverse function of range—the farther a weapon was from its target the less likely it was to strike it. Consequently, despite our images of sharpshooters in warfare, from Robin Hood to Alvin York, it took many, many shots to destroy most of what was being shot at.

Much of the current debate on nuclear weapons emphasizes static strategic principles, the nuclear weapon as the ultimate weapon, and the seeming irrelevance of technological innovation in the face of the awesome destructive power of thermonuclear devices. But the use of force involves the art of risk-taking, and its nature is dynamic, not static. A weapon is ultimate only until its effective countermeasure has been developed and deployed. The only constant in military technology is change.

Military strategy in the nuclear age was not invented in 1945. Many of the essential principles can be found in the classics. Highly relevant to the application of force in the nuclear age is the work of Sun Tzu, written about 500 B.C., during the warring states period.[2] According to Sun Tzu, all warfare is based on deception. The object in war is victory as rapidly as possible; one should avoid protracted warfare at all costs. One should use as little force as necessary to achieve one's goals and take a state intact if possible rather than ruin it. One should attack the enemy's strategy and disrupt his alliances. One should attack the enemy's army and, only last, attack his cities. One should make it impossible for the enemy to know where to attack one. In other words, concealment and deception are essential to remaining invulnerable. In addition, a balance must be struck between predictability, to control one's own forces, and unpredictability, to keep the enemy uncertain about the

2. See Sun Tzu, *The Art of War* (London: Oxford University Press, 1963), pp. 63–149.

intended actions of these forces. Sun Tzu also emphasized the importance of knowing the territory in which conflict takes place and the role of espionage in ascertaining the enemy's strategy and in deceiving him about one's own plans.

Since the time of Sun Tzu the principles of strategy have expanded to include the need for unity of command and the importance of maintaining the initiative, concentrating forces, retaining local superiority regardless of the overall balance of forces, and maneuverability, flexibility, and simplicity. Repeatedly, strategic thinkers have stressed the need for clear objectives and the relating of these objectives to resources available. These principles are just as valid in the nuclear age as in any other. Many of the contemporary strategic issues addressed below—the emphasis on attacking military targets rather than population centers; the stabilizing effects of the largely invulnerable sea-based forces; and the efforts by the Soviet Union to distance the United States from its European allies—can be seen as merely the latest application of these basic ideas.

Capabilities and Trends in the Strategic Forces

For many years the United States has deployed three types of strategic forces—intercontinental ballistic missiles based on land (ICBMs) and at sea (SLBMs) and intercontinental-range bombers—known collectively as the Triad. Additional American nuclear weapons can be delivered by several squadrons of FB-111 fighter bombers stationed in the northeast United States, by carrier-based aircraft in the Mediterranean and the western Pacific, and by ground and air forces stationed in West Germany, South Korea, and elsewhere. But it is the Triad that has carried the burden, as reflected in official public statements, of deterring a nuclear attack by the Soviet Union, of possessing an assured retaliatory capability, of providing flexible targeting and weapon-yield options, and of being in a state of overall equivalence with Soviet strategic forces.

To be sure, the origins of the Triad were not predicated solely on these considerations. Interservice and intraservice rivalry, bargaining between U.S. Secretary of Defense Robert S. McNamara and the Joint Chiefs of Staff in the early 1960s, defense-contractor preferences, and, in the case of the Minuteman land-based missile, the salience of round numbers (that is, the political attractiveness of deploying 1,000 missiles) all influenced the Triad's size and composition. But these factors notwithstanding, the Triad has in fact satisfied American strategic objectives for more than two decades.

The synergistic effect of the three-component force would be absent if one of the components were eliminated. The highly survivable sea-based force and its assured retaliatory capability strengthen the deterrence against a preemptive attack on ICBMs and bombers. Launch of a Soviet ICBM attack on U.S. land-based missiles would give enough warning to allow the bombers to become airborne. And a combined Soviet SLBM-ICBM attack on U.S. bombers and land-based missiles, no matter how configured, would fail to give the Soviet Union an overwhelming military advantage. If the attack were designed so that Soviet ICBMs and SLBMs arrived simultaneously on U.S. targets, the necessarily earlier launching of Soviet ICBMs would immediately alert the U.S. bomber force. If the Soviet forces were launched simultaneously, there would be enough time for U.S. ICBMs to be launched under the Soviet SLBM attack before Soviet ICBMs arrived. (This assumes that U.S. early warning systems function as intended and that political decisions to retaliate can be made in perhaps fifteen minutes, a matter addressed in the next chapter.)

The three components of the U.S. strategic force also provide several different modes of target penetration, vastly complicating the problem of Soviet defense. ICBMs fly high-altitude ballistic trajectories and, because of their yield-accuracy characteristics, are the most effective U.S. weapon for penetrating hardened targets. SLBMs can be fired at close ranges and with depressed (that is, increased energy) trajectories, thereby giving Soviet targets minimum warning time. Manned bombers can penetrate Soviet defenses by flying at extremely low altitudes and have the invaluable advantage of flexibility in target selection, since the pilots can decide where to strike. The Triad, therefore, is a substantial hedge against Soviet breakthroughs in both targeting and defense capabilities.

Historically, elements of the Triad have been intended for use against all potential Soviet targets: population centers, known as *countervalue* targets; the war-making capacity of the Soviet state, known as *urban-industrial* targets; and the Soviet military forces, known as *counterforce* targets. A subset of targets—command posts, certain factories, and missile silos, for example—that are protected by reinforced concrete and other means to better withstand the effects of nuclear blast, radiation, and electromagnetic pulse are called *hardened* targets. Targets that need to be struck quickly before their value diminishes (for example, missile silos before missile reloading) are *time-urgent* targets.

While there is nothing magical or preordained about the Triad, its retention has made good strategic sense. Therefore, when considering force posture issues and alternatives the interdependencies of each U.S. strategic force component, as well as its strengths and weaknesses, must be kept in mind.

Of course, Soviet forces and American perceptions of their effectiveness are important determinants of U.S. defense. Official U.S. estimates of the quantitative and qualitative attributes of these forces, projections of the pace of their deployment and modernization, and overall assessments of the U.S.-Soviet strategic balance justify American weapon deployments and research and development programs.

Much of the information about Soviet weapon systems is collected by reconnaissance satellites that carry high-resolution photographic and other types of equipment. These satellites, which were first deployed in the early 1960s, can reveal fine details of uncamouflaged fixed-site weapon systems, missile silos, and airfields. In addition, electro-optical sensors on board ships and aircraft and land-based radars are used to monitor Soviet missile tests and radar operations, and a variety of land- and sea-based listening posts attempt to intercept Soviet communications.

Although much information is consequently available about forces in place, there is a great deal of uncertainty about the performance characteristics of these forces and future trends in Soviet weapon development. The SALT agreements have not greatly reduced this uncertainty, except with respect to antiballistic missile systems; it remains a principal cause for the caution and conservatism that characterize U.S. strategic force planning.

Since the mid-1960s, perhaps stimulated by the humiliating outcome of the Cuban missile crisis or by an incipient Sino-American alliance, Soviet leaders have invested enough resources in their strategic forces to greatly increase and improve their capabilities. The United States now knows that in the early 1960s the Soviet Union possessed only two dozen ICBMs, compared with the several hundred U.S. ICBMs, and that the performance of Soviet systems lagged appreciably behind that of their American counterparts. Moreover, Soviet systems were very vulnerable to attack and could not be launched quickly. The few years of the early 1960s, then, constituted a brief period when the United States could perhaps have carried out a disarming strike against Soviet forces. Neither side has possessed this capability since. By the early 1980s the Soviet Union had surpassed the United States in numbers of weapons and had markedly improved weapon system performance.

Although there are many pronounced asymmetries in the strategic forces of the two states, there are also many common features that form a framework for comparison.

Intercontinental Ballistic Missiles

Both superpowers have invested heavily in a fleet of sophisticated fixed-based ICBMs, in many ways still the showpiece of each force. Roughly 75 percent of total Soviet megatonnage (a measure of explosive power) is contained in the Soviet ICBM force, whereas only about 25 percent of U.S. megatonnage resides in ICBMs. Both sides have deployed multiple independently targetable reentry vehicles (MIRVs) on many ICBMs, thereby enabling a single delivery vehicle to strike several different targets in a single attack. The United States has more solid-fueled than liquid-fueled missiles. The former are more reliable and can be launched more quickly than the latter.

Land-based missile forces have many strengths. Networks of hardened, underground facilities provide relatively secure and effective means of command, control, and communication. The basing of the ICBMs in fixed sites on U.S. and Soviet territory facilitates communication between the national command authorities and among the various ICBM launch-control centers. In contrast, communication with deeply submerged submarines carrying ballistic missiles is extremely difficult. Both U.S. and Soviet ICBM forces carry sophisticated guidance packages that allow great accuracy in the delivery of warheads on time-urgent targets. Most important, the U.S. and Soviet ability to destroy hardened targets now resides in their ICBM forces. U.S. ICBMs and perhaps some Soviet ICBMs also have retargeting flexibility not characteristic of sea-based forces. In addition, fixed-site ICBMs have relatively low operational costs and are thought to be extremely reliable. The notable weakness of land-based missiles—primary symbols of military power that have political uses in relations with both adversaries and allies—is their vulnerability, while based in fixed silos, to counterforce attacks. It was widely accepted in the early 1980s that in this area the Soviets have an advantage. According to theoretical calculations, the Soviets can launch roughly 30 percent of their ICBM force and destroy more than 90 percent of the U.S. ICBMs in fixed silos, a performance capability that U.S. forces cannot, as of 1984, duplicate.[3] The military utility of this Soviet advantage is a point of contention. Given the large number of warheads deployed on other systems, some analysts are not overconcerned about this asymmetry. Nonetheless, the condition has taken on enormous political importance in the United States because it feeds on chronic American fears, since Pearl Harbor, of surprise attack and because it is partial evidence that America is "losing the arms race."

Sea-Based Ballistic Missiles

The sea-based forces are the least vulnerable element of the U.S. Triad. U.S. submarine-launched ballistic missiles (SLBMs) are deployed on a fleet of nuclear-powered ballistic missile submarines (SSBNs) that, when on station, encircle the Soviet Union. Both U.S. and Soviet programs are moving in the same general direction. Both countries are developing SLBMs with MIRVs of increasing accuracy; longer-range SLBMs that permit the SSBNs to expand their patrol areas to millions of square miles, thereby reducing their vulnerability to antisubmarine warfare (ASW) attack; and deployment of quieter SSBNs that can stay submerged at greater depths for longer periods and maintain faster cruising speeds than their predecessors. In time, the SLBMs of both countries will probably acquire the hard-target kill capability now attributable to the ICBMs. If this occurs, it would provide the strategic logic, though not the organizational incentives, to phase out the fixed land-based missiles of both the U.S. and Soviet strategic forces.

One apparent error of both forces is the deployment of larger SSBNs. As long as both sides adhere to arms control limitations established in the SALT I Interim Agreement and the SALT II Treaty, this has the effect of placing more eggs in fewer baskets—that is, more missiles on fewer submarines—thereby increasing the vulnerability of both forces.

Most experts agree that the United States possesses important advantages. The accuracy and reliability of Soviet systems are considered to be substantially inferior to the frontline U.S. SLBM systems. Soviet

3. See U.S. Department of Defense, *Soviet Military Power, 1984* (Government Printing Office, 1984), p. 23.

strategic submarines are believed to be much noisier and therefore potentially more vulnerable than their U.S. counterparts. The Soviets also need more time to overhaul a nuclear submarine than do the Americans. The result is that only about 20 percent of Soviet frontline SSBNs are away from port at a given time, compared with 50 percent of American submarines.

The Soviets also have a major geographic disadvantage. They have limited access to warm-water ports, and their points of access to the open ocean can be much more easily monitored by U.S. ASW forces than U.S. SSBNs can be monitored by Soviet forces. The Soviets are, however, deploying longer-range SLBMs, which will partly negate this disadvantage. Finally, the United States allegedly is ahead of the Soviet Union in ASW technology, although there is limited information publicly available to verify this assertion.[4]

Strategic Bombers

Intercontinental-range bombers are the most flexible of the strategic systems because they can be recalled and because targets can be reset according to pilots' discretion. The effectiveness of bombers in a nuclear war is uncertain, however. Intercontinental-range bombers remain the least developed element of the Soviet strategic offensive forces. For more than a decade there has been no significant change in the composition of the Soviet bomber force, although U.S. intelligence analysts have expected a new advanced bomber for several years.[5]

The United States is most concerned about the ability of its bombers to survive the warning-to-launch period and the ability to penetrate Soviet defenses. The former will become more precarious if the Soviets develop depressed-trajectory SLBMs that could destroy U.S. bombers while still on the ground. The latter problem will increase as the Soviets develop an effective way to detect, track, and attack low-altitude U.S. targets. In the meantime the task of American bombers is to penetrate a complex network of more than 12,000 surface-to-air missile systems and

4. A useful discussion of this subject is Richard L. Garwin, "Will Strategic Submarines Be Vulnerable?" *International Security,* vol. 8 (Fall 1983), pp. 52–67.

5. This discussion is limited to Soviet bombers with estimated maximum ranges, without refueling, of more than 6,000 nautical miles. Such aircraft are also sometimes called heavy bombers. A new Soviet bomber, the Blackjack, may become operational in the late 1980s.

hundreds of fighter interceptors to reach their targets. This can be done by using cruise missiles to penetrate the defenses while the bomber "stands off" Soviet territory or by using a combination of cruise missiles and short-range attack missiles (SRAMs) to create "bomber-penetration corridors" once sections of the Soviet air defenses have been destroyed. A third technique being developed is the application of new materials that would dramatically reduce the radar cross section (the profile "seen" by the defense) of the attacking aircraft and thus greatly increase their ability to penetrate defenses. These "Stealth" aircraft may be deployed in the early 1990s; in the interim, it is now anticipated that the United States will deploy the B-1B manned bomber equipped with Stealth technology.

Gray-Area Systems

In addition to ICBMs, SLBMs, and strategic bombers, both the United States and the Soviet Union are developing and deploying weapon systems that are of intermediate range but that in some cases can deliver nuclear warheads to the homeland of the adversary. These have been labeled gray-area systems because their capabilities further blur the already artificial distinction between strategic and tactical weapons.

Traditionally, the term *strategic* has referred to weapons designed to strike at the source of an enemy's military, economic, and political power; the term *tactical* has referred to weapons primarily for use on the battlefield. But this differentiation does not always hold true, and the term *strategic* has taken on a variety of meanings.[6]

Four gray-area weapon systems are noteworthy: the Backfire bomber, the SS-20, the Pershing II IRBM, and the long-range cruise missile. The Backfire is a supersonic bomber of variable wing geometry that the Soviets first deployed in 1974. Initially it was classified in the West as a theater-weapon system, designed for use against targets in Western Europe and China and to provide shore-based air support for Soviet naval operations. These uses would classify it as a tactical weapon

6. Henry S. Rowen has pointed out that the term has referred to "(1) attack by United States or Soviet forces on opposing homelands; (2) attack on population (and/or industry) as distinct from military targets; (3) attack on missiles in silos and other long-range forces versus attack on general purpose forces; (4) attack on 'deep' targets; (5) nuclear as opposed to non-nuclear attack; (6) attacks using long-range vehicles against any target; or (7) any attack launched from outside the theater." See Rowen, "The Need for a New Analytical Framework," *International Security*, vol. 1 (Fall 1976), pp. 138–39.

system, similar to U.S. forward-based aircraft, and thus not part of the SALT agreements and estimates of the Soviet-American strategic nuclear balance. However, analyses of the Backfire's performance characteristics led to the conclusion that, if the aircraft were based in the Chukchi Peninsula of northeast Siberia, it would be capable—on a round-trip, subsonic flight without refueling—of reaching targets along an arc from Los Angeles to the western tip of Lake Superior. Midair refueling would permit two-way missions to include most significant targets in the United States.[7]

The SS-20 is a medium-range, two-stage, solid-fuel ballistic missile with three independently targeted warheads that the Soviets are deploying in the western part of the USSR and near the Sino-Soviet border. By replacing the firing cannister in which the SS-20 is housed with a larger one and by adding a third stage, the missile could be given enough range to strike various targets in the United States. Primarily, however, the missile is strengthening the Soviet Union's position of nuclear superiority in the European theater. What disturbs the Soviets is that NATO began in late 1983 to deploy in West Germany Pershing II intermediate-range ballistic missiles, which could strike the Soviet homeland (particularly command centers that control Soviet strategic forces) with only a few minutes' warning time. Although the United States sees the Pershing II missile as a theater—that is, intermediate-range—weapon, the Soviet Union sees it as a strategic weapon.

Finally, both superpowers are developing and have begun to deploy long-range cruise missiles that can be based on land, at sea, or in the air. Cruise missiles are pilotless aerodynamic vehicles that are propelled continuously from launch to impact. Air-launched cruise missile systems designed primarily for battlefield use, such as the Soviet Kangaroo and the American Hound Dog, have been deployed for several years. But advances in propulsion efficiency and guidance-system technology have made it possible to deploy a new class of weapons: small (perhaps less than twenty feet long and twenty inches across), highly accurate, subsonic cruise missiles, which can be launched from the air, sea, and land and which can hit targets at ranges of up to 2,000 nautical miles. Advanced cruise missiles integrate small warheads, sophisticated guidance technology, and highly efficient propulsion systems into extraordinarily accurate and reliable weapon systems.

Many arguments support widespread deployment of cruise missiles

7. *United States Military Posture for FY 1978,* p. 20.

in the U.S. strategic arsenal. Air-launched cruise missiles compensate for the anticipated decline in effectiveness of the penetrating bomber force by allowing the Triad to strike targets deep in Soviet territory. Deployment of long-range cruise missiles at sea, under the sea, in the air, and on the ground would greatly complicate Soviet air defenses and at the same time would frustrate Soviet efforts to acquire a disarming counterforce capability against all U.S. strategic forces. Because cruise missiles can be deployed with nuclear and conventional warheads and have long-range as well as theater applications, they would greatly enhance the flexibility of American forces.

Consequently all of the military services are interested in cruise missiles, although this was not initially true. Doctrinally, cruise missiles are appealing because their subsonic flight speeds and multiple basing make them fine second-strike (but not first-strike) weapons.[8] Moreover, cruise missiles combine efficient propulsion systems, small warheads, and sophisticated guidance systems—a "technologically sweet" package that symbolizes American technological sophistication and therefore has political and economic appeal.

Cruise missiles are attractive to some allied governments as well. They are a possible successor or supplement to the British sea-based deterrent; they are a probable addition to French tactical and strategic nuclear forces; and they would enhance West Germany's ability to fight a war against the Soviet Union, an admittedly mixed blessing from the German perspective. Indeed, it is for precisely these reasons that the Soviet Union sought strenuously to prevent their deployment. Cruise missiles have the potential to degrade its air defenses, diminish its military advantage in relation to Western Europe, and eventually enhance the military capabilities of China and other states that might be in conflict with Soviet interests. The principal argument against the deployment of cruise missiles is that they could undermine or even destroy future prospects for negotiated arms control agreements on offensive strategic forces.[9]

8. Because they are not first-strike weapons they would not provide a political leader the incentive to strike first in a crisis. But they may greatly stimulate the arms competition. Therefore, it can be argued they enhance "crisis stability" but undermine "arms race stability." Should supersonic or hypersonic cruise missiles be developed, they would no longer serve crisis stability.

9. The most comprehensive account of the implications of cruise missile deployments is Richard K. Betts, ed., *Cruise Missiles: Technology, Strategy, Politics* (Brookings Institution, 1981).

The Importance and Implications of High Accuracy

Throughout the more than thirty-year evolution of strategic weapons, there have been few truly critical developments. In the early 1950s classified work done for the U.S. Air Force at the Rand Corporation developed fully the notion of a second strike, which called for the deployment of a nuclear force that could survive an initial nuclear attack and be able to respond selectively against military targets. The purpose was to create an effective nuclear deterrent. The Rand study focused on

> how to choose a protected mode of basing and operating [of] a strategic force that would be best for any of several target systems. It looked at several target sets typical of the time: a quite limited number of key war plants supporting combat; military targets whose destruction might retard the advance of ground forces in Europe; and those that might blunt a continuing enemy strategic attack. It did so in order to show in all cases how best to reduce the vulnerability of our own strategic forces.[10]

The establishment of criteria for a retaliatory force was the first major conceptual development of the nuclear age. Then, with the marriage of nuclear weapons, ballistic missiles, and electronic guidance systems, technology took over. American as well as Soviet cities and "soft" military and industrial targets in both countries became vulnerable to nuclear annihilation. But after the Soviets had dispersed and begun to harden their land-based nuclear forces in the mid-1960s, neither side could claim the ability to launch a disarming counterforce attack against the second-strike forces of the other side. At least temporarily, a state of crisis stability had been achieved, since neither side was willing to unleash its nuclear arsenal even in a situation of great stress, for fear of the damage that could have been inflicted by second-strike forces. From this time on, then, the principal guideline for strategic force planning and procurement has been this strategic counterforce ratio: the number of offensive weapons in a first strike necessary to destroy a retaliatory weapon and, conversely, the number of retaliatory weapons that could be destroyed by an offensive weapon in a first strike.

The next critical development was that of antiballistic missile systems, although they were never fully deployed. Both Soviet and U.S. ABM systems could perhaps have been significant in the strategic balance of

10. Albert Wohlstetter, "Bishops, Statesmen, and Other Strategists on the Bombing of Innocents," *Commentary*, vol. 75 (June 1983), p. 21.

the 1970s if they had been fully deployed to defend cities and ICBM fields. But the ABM Treaty (part of SALT I) of October 1972 limited each side's deployment to very modest numbers of ABM launchers, interceptors, and radars. Both countries had many complex reasons for making this agreement (they are examined later in the chapter on arms control), but they no doubt included—at least for the United States—the technical deficiencies of the system. The system's radars were vulnerable to attack; the system could not cope with large attacks that would saturate its interceptors; the system's target discrimination capability could be degraded by decoys, chaff, and other penetration aids; and the system could too easily leak (that is, permit attacking warheads to reach their target) under a sophisticated attack.

With each side defenseless against a nuclear attack, efforts shifted to improve offensive capabilities, especially by deploying MIRVs on ICBMs and SLBMs. Both sides continued this practice throughout the 1970s, a practice that has highlighted the difficulties likely to materialize in an age of high-accuracy weapons. Conceptually, the worrisome characteristic of MIRVs is precisely what makes them so attractive strategically: several warheads on a relatively small number of launchers—a large warhead-to-launcher ratio—permits an attacker to expend a small percentage of his force to destroy a large percentage of the other side's force. (This is assuming that the number of the attacker's warheads with MIRVs is greater than the number of launchers of the targeted force, which is the case for both countries in the current strategic balance.) Consequently the incentive to strike first rises dramatically as the accuracy of reentry vehicles improves and as the location of the adversary's force becomes known.

Most of all, high accuracy of MIRVs highlights the vulnerability of land-based missiles. This problem has probably been studied more intensely over a longer period than any other issue in modern U.S. defense policy. There is no particular reason to review the detailed solutions that have been proposed and then rejected.[11] However, it is useful to understand the policy problem and a possible conceptual route out of it.

It was recognized as early as the early 1950s that fixed-based systems would eventually be vulnerable to missile attack. But only in the early 1970s did it become apparent that the heightened accuracy of Soviet

11. An assessment of eleven different basing modes is presented in Office of Technology Assessment, *MX Missile Basing,* OTA-ISC-140 (GPO, 1981).

MIRV-equipped ICBMs and the projected effectiveness of these weapons were a serious threat to the U.S. Minuteman force.

Analytically, this effectiveness can be computed as follows:

$$P_k = 1 - e^{-cY^{\frac{2}{3}}/H^{\frac{2}{3}}(CEP)^2},$$

where

P_k = the single-shot kill probability, the probability that a target will be destroyed by a single attacking warhead

Y = the yield (energy equivalent to tons of TNT) of the attacking warhead, expressed in kilotons or megatons

H = the hardness of the target, that is, the pressure above normal atmospheric pressure needed to destroy the target, expressed in pounds per square inch (psi)

CEP = the circular error probable, that is, the radius, expressed in nautical miles or feet, around the target within which the attacking warhead has an 0.5 probability of landing

c = a constant dependent on the units of measurement.

Because Y is raised to the two-thirds power while CEP is squared, the single-shot kill probability is far more sensitive to increases in warhead accuracy than to increases in warhead yield. Moreover, since H is raised to the two-thirds power, accuracy improvements can overcome increased silo hardening.

According to estimates based on observed test data, Soviet ICBMs have achieved accuracies of 0.1 CEP, and these accuracies are expected to improve still further for Soviet land-based and sea-based forces throughout the decade. Although the United States is increasing the hardness of its missile silos to withstand pressures of perhaps 3,000 pounds per square inch or more, a Soviet attack on each silo using two reentry vehicles with warhead yields of one megaton and accuracies of 0.1 CEP or better would nonetheless result in kill probabilities exceeding 90 percent.[12]

Theoretical calculations of kill probabilities, however, do not take into account important operational considerations that could strongly influence the success of an attack on hardened targets. For example, a

12. Since fiscal 1983 research has been conducted on the feasibility of constructing "superhard silos" able to resist overpressures in excess of 15,000 psi. If achievable, the demands on the attacking Soviet force would be increased considerably. See Edgar Ulsamen, "The Prospect for Superhard Silos," *Air Force Magazine*, vol. 67 (January 1984), pp. 74–77.

multiple warhead attack on a single silo could be subject to ''fratricide'' effects, in which the detonation of one attacking warhead could produce an environment near the target turbulent enough to deflect or destroy subsequent warheads. Nuclear radiation, shock waves, vast amounts of dust and debris, winds of great velocity, very strong oscillating electrical and magnetic fields, and the nuclear cloud resulting from the first warhead detonation not only could degrade the performance of subsequent warheads attacking the same silo, but could also deflect or disable attacking warheads aimed at adjacent silos.

The perceived vulnerability of U.S. land-based missiles depends, therefore, on assumptions about the launch reliability and the flight reliability of the attacking forces, possible interference or fratricide effects, the number of warheads attacking each silo, the way in which the attack is timed, climatic conditions over the target area at the time of attack, and the values in the single-shot kill probability formulas.[13] In addition, assumptions need to be made about whether the Soviet Union would use SLBMs to pin down the land-based missiles in their silos until Soviet ICBMs had arrived and whether the United States would or could launch some or all of its ICBMs while under attack.

Another uncertainty is the projected number of civilian casualties that would result from an attack on American fixed-site ICBMs. It is argued that a Soviet strike against the Minuteman force is less likely to provoke American retaliation against Soviet cities, and will hence be a more credible Soviet option, if the initial attack does not kill too many American civilians. But there are substantial differences of opinion about the potential number of fatalities; estimates range from 800,000 to 22 million.[14]

Faced with these uncertainties, successive administrations since that of President Nixon have grappled with the problem of fixed-site missile vulnerability. The air force proposed the deployment of the MX missile

13. Considerations related to the kill probability formulas that complicate matters still further include the oversimplification of the notion of a single hardness value for a given target, the effect of blast wave duration on silo vulnerability, the uncertainty of the CEP values ascribed to attacking warheads, and the dependency of the kill probability formula on the concept of a lethal radius that is tractable analytically but only approximates the physical reality. These considerations are discussed in Michael L. Nacht, ''The Vladivostok Accord and American Technological Options,'' *Survival,* vol. 17 (May–June 1975), pp. 109–10.

14. See *Effects of Limited Nuclear Warfare,* Hearing before the Subcommittee on Arms Control, International Organizations and Security Agreements of the Senate Committee on Foreign Relations, 94 Cong. 1 sess. (Government Printing Office, 1976).

as the fourth generation U.S. ICBM (following Atlas, Titan, and various versions of Minuteman) and has been seeking a mobile basing mode for its deployment. Scores of different basing modes have been studied in detail, but each has been rejected on technical, economic, or domestic political grounds. Having failed to find a technically and politically acceptable mobile basing mode, the Reagan administration plans to deploy a limited number of MX missiles in Minuteman silos, even though the MX would be vulnerable to a Soviet ICBM attack using MIRVs.

The decision has been defended on the grounds that this deployment would show resolve, facilitate strategic arms reduction negotiations with the Soviet Union, and be coupled with the development and subsequent deployment of a new, small, mobile, single-warhead missile that would contribute to strategic stability.[15] Since it is not at all clear that arms reduction negotiations will be facilitated by deploying MX missiles in Minuteman silos and since the small missile is still in the earliest stages of research, the decision in the short term may amount to deploying a highly valued weapon—the MX missile—in highly vulnerable silos. For the missiles to retain their value, they would have to be used before they were attacked.

Since large numbers of high-accuracy weapons are in the offing, deceptive basing schemes are undesirable because they can be overwhelmed by saturation attacks against all fixed targets. Mobile systems are desirable but remain vulnerable to barrage attacks against the entire basing area. The most preferable solution to the ICBM-vulnerability problem is to combine base mobility with concealment, which is the basis of SSBN invulnerability. However, moving to the sea makes U.S. missiles susceptible to antisubmarine warfare. Such a move concentrates U.S. targets for Soviet counterforce attacks rather than complicates Soviet planning. Air-mobile missile basing is another possible solution, but aircraft need airfields, which are totally vulnerable fixed targets. The only aircraft that do not need airfields are short takeoff and landing (STOL) amphibious aircraft, which can move continuously and land randomly on the high seas, in coastal waters, in inland waterways, or on land. Such aircraft could carry one or two ICBMs, each about half the size and weight of the MX. The missiles could be housed in cannisters attached to the outside of the aircraft hull. The cannisters would be released from the aircraft and become vertical with the aid of a drogue

15. See "Report of the President's Commission on Strategic Forces" (Washington, D.C., April 1983).

chute. A set of parachutes would reduce the cannisters' descent rate, and then the missiles would be fired.

A fleet of 100 aircraft carrying two missiles each would be a potent retaliatory force and present the Soviets with an active trailing problem of gigantic proportions. Basing missiles at sea would eliminate the U.S. domestic political opposition that has long plagued the MX basing issue. The technical risk of using STOL amphibious aircraft as missile bases is minimal because the aircraft could be designed from state-of-the-art technology and the new missiles they would carry could benefit directly from the MX research and development program. To achieve high accuracy the missiles would need position updates, but the global positioning system could provide this capability in due course. Moreover, the system would be impervious to the growth of the Soviet threat. Command and control presents no greater a problem than it does for the SSBN force. And just as the number of deployed SSBNs can be identified for purposes of verifying compliance with arms control agreements, the same conditions would hold for a finite number of these amphibious missile platforms.

There are three potential drawbacks to this potential solution: cost, endurance, and organizational complexity. Procurement costs would be higher than for land-based alternatives because it is much more expensive to procure aircraft than it is to build silos. Operating and maintenance costs would also be high. Endurance—the capability to survive as an integrated system for a long period after a nuclear attack—would be low, but no lower than the endurance of the SSBN fleet. Finally, it is not clear whether the air force, the navy, or some combined force would man the fleet of aircraft, a matter certain to arouse intense interservice rivalry, although perhaps no more intense than that experienced in the organization of the Rapid Deployment Joint Task Force, or "Central Command."

This missile deployment mode, which the Department of Defense considered briefly and rejected, is no mere "Rube Goldberg" solution despite its unorthodox characteristics. Indeed, it is exactly this sort of innovative force planning that should be considered in an age of high-accuracy systems. Moreover, in thinking about what constitutes a credible retaliatory capability, it is important to remember the lack of automaticity involved. The retaliatory forces must be able to survive initial enemy attacks, be provided the command to retaliate, reach enemy territory, penetrate active and passive defenses, and only then destroy the target. It is not enough to look solely at the offensive side of the equation when judging the effectiveness of nuclear retaliatory forces. Indeed, assuming the United States deploys weapons with some form of prompt counterforce-targeting capability on land, at sea, or in the air, a comparable problem will materialize for the Soviets, who have deployed most of their nuclear forces in fixed, and therefore increasingly vulnerable, silos. For the United States it makes most sense to augment the existing Triad and cruise missile force with a fleet of ballistic missiles based on amphibious aircraft. Procurement of neither the MX nor an arsenal of single-warhead missiles would then be warranted. The amphibious alternative, and no other, would enhance ballistic missile survivability, be easily embraced within the existing arms control regime, and meet the powerful American domestic political constraints that have plagued this issue for a decade.[16]

ICBM vulnerability illustrates the consequences of high accuracy in the nuclear battlefield. The time is approaching when most shots fired will hit their targets, assuming the new systems perform as advertised. The world is moving into a new era of cruise missiles with terminal guidance provided by terrain-matching capabilities or satellite-based positioning systems. Scores of television-guided, laser-guided, and optically guided weapons will allow the user to both hit what he wants to hit and not hit what he doesn't want to hit. The implications of this accuracy can be summarized as follows.

—Fixed targets will become increasingly vulnerable, and thus there will be a far greater premium on maneuverability, concealment, and deception.

—Accuracy will no longer be a function just of range. Long-range intervention will become more feasible, and "remote" means will no longer be remote.

—The clustering of forces will become more hazardous, and the dispersal of units will tend to increase.

—The collateral damage—that is, damage to areas other than the

16. The essential weakness of the concept is that it has no natural organizational constituency. The Office of the Secretary of Defense should authorize the conduct of systems analysis and engineering feasibility studies to determine the concept's operational strengths and weaknesses. If the system satisfies appropriate technical and cost-effectiveness criteria, a suitable air force–navy mixed-service arrangement must be found for program implementation to succeed. This is one case in which the attractiveness of the concept should drive organizational considerations rather than permit entrenched vested interests to produce undesirable strategic policies.

target—resulting from the effective use of high-accuracy weapons is likely to be low.

—Logistics facilities and second-echelon forces in rear areas will become more vulnerable.

—The high accuracy of weapon systems will permit reductions in warhead yield. In particular, tactical nuclear weapons could be replaced by conventional warheads.

In general, the acquisition of a variety of high-accuracy systems will raise the nuclear threshold. In the event of war, military commanders and political leaders could have their cake and eat it, too: there would be a high probability of destroying the target without resorting to the use of nuclear weapons.[17]

The Revitalization of Active Defenses

Ironically, the most significant impact of advanced high-accuracy systems on the U.S.-Soviet strategic nuclear balance will probably be the reactivation of active defenses. Active defenses—ballistic missile defenses (BMD), ASW systems, and air defenses—seek to destroy attacking forces before they reach their targets.[18] The United States has access at present to two categories of technology, neither one particularly advanced, that could protect U.S. ICBMs. First, there are novel methods using unsophisticated components to create uncertainty for the attacker but that are also unreliable for the defender. Second, the United States has access to components that are technological derivatives of programs from the early 1970s, especially the low-altitude defense system (LOADS). Despite the intense public debate over "star wars"—a popular euphemism for directed-energy missile defenses in space, proposed by President Reagan in 1983—it is these relatively mundane technologies that are now within the range of feasibility.[19]

17. One system already in engineering development for use in the European theater is "Assault Breaker," a missile that homes in on a target and then releases smaller missiles using conventional warheads on individual targets. See Flora Lewis, "New Nuclear Vision," *New York Times,* October 8, 1982.

18. Passive or civil defenses are designed to absorb the destructive power of the attack so as to limit significantly the damage to the intended target. They are discussed in chapter 5.

19. In this speech President Reagan called for "the means of rendering these nuclear weapons impotent and obsolete." See "President's Speech on Military Spending and a New Defense," *New York Times,* March 24, 1983.

The first category includes such methods as "beds of nails" —ground-based metal spikes that could be fired at incoming warheads. Theoretically, the warheads would be destroyed before they could detonate against the missile silos. In reality, however, the effectiveness of this method and several others incorporating similar technology would depend heavily on the specific nature of the incoming attack. Moreover, for some of these methods to be tested the ABM Treaty would have to be renegotiated, since the treaty prohibits the testing of mobile land-based BMD systems or systems with multiple launchers or rapid reload capability.

In any event, because these novel systems are not very reliable, the BMD technical community has concentrated on the LOADS, at least in the short term. It is an advanced form of terminal defense involving the use of small radars, modern data processing systems, and a single-stage ABM interceptor that would destroy attacking warheads between 50,000 and 200,000 feet above the ground when they are within 10 seconds of their targets. In a sense the LOADS is high risk because it is designed to intercept the incoming warhead at the last possible moment. If the interception is not made at low altitude as planned, the defense has no second chance.

The LOADS would be particularly attractive in defending MX missiles deployed in a multiple protective shelter (MPS) system, in which one MX missile would be hidden within a cluster of shelters. In such a scheme each LOADS interceptor would defend one shelter in each cluster—the one with the MX. The LOADS unit itself would be housed within one of the shelters, and the attacker would not know which shelter the unit was defending. The attacker would consequently have to aim a warhead at each shelter in the cluster to be sure of destroying the single MX missile.

Because the essential feature of this system is its deceptive basing, however, it would violate the ABM Treaty. Thus, deployment of the LOADS would require renegotiating the treaty. Moreover, the system's radars are vulnerable to jamming; the system has not been designed to counter maneuverable reentry vehicles, which the Soviet Union may be able to deploy by the late 1980s or the early 1990s; and the entire basing concept is contingent on the Soviet Union's inability to determine quickly which shelter is hiding the missile.

The effectiveness of the LOADS declines dramatically when it is not paired with a deceptive basing scheme. If LOADS units were deployed to defend fixed missile silos, the Russians would have to aim only one additional warhead at each missile site to be sure of destroying the missile, assuming the LOADS units themselves were not deceptively based to introduce uncertainty about how many interceptors were defending each ICBM. The United States would, therefore, have trouble justifying the deployment of the LOADS for fixed-based MX missiles. Nonetheless, for the next decade a LOADS-type system for limited terminal defense is the only feasible option. Given its inherent deficiencies, its adverse consequences for the ABM Treaty, and its stimulus to offensive systems, there is no basis at present to warrant its deployment.

Looking further ahead, layered BMD systems, combining an exoatmospheric kill capability with a low-altitude BMD system like the LOADS, may be available within twenty years. This technique, which would use nonnuclear interceptor missiles, equipped with MIRVS, could reduce the number of attacking warheads that reach their targets. Testing or deployment of layered BMD systems would also require revising the ABM Treaty.

The adequacy of defense against nuclear attack in the future depends, of course, on the state of weapon technology and new technological developments that could radically change the strategic balance of forces. The practical question is whether the new technologies will make feasible a shift to a defense-dominated strategic balance. A more ominous question is whether either the Soviet Union or the United States would acquire the capability to launch a disarming first strike against the retaliatory forces of the opponent.

There are now three areas of research and development that have revolutionary rather than merely evolutionary potential: directed-energy weapons, the exploitation of space as a theater of warfare, and a third generation of nuclear weapons. In the aftermath of his "star wars" speech, President Reagan has called for a multibillion-dollar strategic defense initiative (SDI) as a long-term research and development program to assess the feasibility of these technologies.

Directed-energy weapons are high-energy laser and charged-particle-beam weapons that could serve either as effective defenses against attacking ballistic missiles or as antisatellite weapons. The U.S. Department of Defense has maintained limited research programs in both areas since the late 1950s but at low funding levels and with primarily endoatmospheric applications in mind. The revelation in 1977 that the Soviet Union had an active program in both fields at a test facility at

Semipalatinsk, however, stimulated great interest within the U.S. defense community about the potential of these technologies, although professionals were strongly divided at the time about their feasibility.[20]

A laser (light amplification by stimulated emission of radiation) is a highly concentrated beam that uses the natural oscillations of atoms to amplify or generate electromagnetic waves in the visible region of the light spectrum. The laser radiation can be stimulated chemically, by electric discharge, by means of a solar pump, or by an electron beam—four among many techniques. In laser weapons, molecules of gas excited into a higher energy state by one of these techniques would emit photons (quantums of radiant energy), which would collide with adjacent molecules and stimulate more photon emissions. Mirrors would then collect the photons and focus and aim the beam. Target tracking, aiming, and reaiming capabilities would all have to be integrated with the laser beam to form a coherent weapon system. The problem with laser weapons is that they would require enormous amounts of power and would be vulnerable to countermeasures, such as decoys, reflective screens, and target rotations that spread the laser's heat, or to direct attacks.

The Reagan administration, under the SDI, is trying to secure major budgetary support for several long-range laser-weapon programs. The defense community supports two ideas. One is the maintenance of several hundred lasers around the country, each operating at or near the visible light spectrum. The lasers would be fired at huge, flexible mirrors that would themselves be launched upon warning of a potential attack. The mirrors would then refocus the beams onto the attacking booster rockets. The second idea is for hundreds of lasers to be positioned atop missiles and launched into space on warning of an attack. The lasers would be powered by low-yield nuclear explosions and, when detonated, their radiation would allegedly destroy attacking Soviet boosters shortly after launch.[21]

Directed-energy weapon technology is clearly in its infancy. It will be several years before the technology is advanced enough to make serious

20. See *Aviation Week and Space Technology,* vol. 105 (August 30, 1976), p. 14, and particularly *Aviation Week and Space Technology,* vol. 106 (May 2, 1977), pp. 16–23.

21. R. Jeffrey Smith, "The Search for a Nuclear Sanctuary (I)," *Science,* vol. 221 (July 1, 1983), pp. 30–32. An exceedingly detailed and persuasive technical critique of these concepts is Ashton B. Carter, *Directed Energy Missile Defense in Space: Background Paper,* OTA-BP-ISC-26 (Washington, D.C.: Office of Technology Assessment, April 1984).

calculations about its cost-effectiveness. But conceptually, the image of a highly concentrated beam traveling at the speed of light and destroying targets thousands of miles from its point of origin virtually instantaneously is too alluring for the technical community to resist. The task now is to create, by using intense lasers or charged-particle beams, an effective weapon system that is survivable, reliable, and not easily countered.

Charged-particle beams are in an even earlier stage of development than laser weapons. A charged-particle beam focuses and projects atomic particles at the speed of light. As a weapon the beam could, in theory, either be directed from ground-based sites into space to intercept and destroy reentry vehicles or be satellite-based and designed to destroy ICBMs or SLBMs during the boost phase of their flight.

The potential effectiveness of charged-particle-beam weapons is severely limited, however. Unlike a nuclear-armed BMD interceptor, a charged-particle-beam weapon (like a laser weapon) must strike a reentry vehicle to injure it. This means that the weapon's radar accuracy and beam-pointing accuracy must be perhaps three times as great as that of a BMD interceptor. Moreover, the charged-particle beam is bent substantially by the earth's magnetic field, so that it cannot be aimed directly at its target. The attacker can use nuclear explosions outside the atmosphere to disturb the earth's magnetic field and deflect the path of the weapon. In addition, there are serious doubts about whether primary power sources would be able to produce enough energy in extremely short periods of time to make operation of the weapon feasible.[22] The difficulties in perfecting the weapon and the possible countermeasures to it—not only distortion of the magnetic field but also use of decoys and saturation techniques—suggest some of the development problems ahead. Moreover, development of advanced-technology BMD systems is prohibited under the ABM Treaty. As long as the treaty is upheld, full-fledged application of a particle-beam system cannot begin.

Virtually no detailed information is publicly available on either the U.S. or Soviet directed-energy programs.[23] But despite misgivings about the merits of these programs, budgetary support for them and interest in them have risen markedly following President Reagan's endorsement of

22. See Richard L. Garwin, "Charged-Particle Beam Weapons?" *Bulletin of the Atomic Scientists*, vol. 34 (October 1978), pp. 24–27.

23. See "Directed Energy Programs," in *Fiscal Year 1983 Arms Control Impact Statements* (GPO, March 1982), pp. 298–332, for the barest outline of the U.S. program.

the SDI. It should be realized from my earlier discussion of the American national character just how powerful the appeal of the SDI is: it is a technological fix to an enormously complex set of operational problems; it calls for defense rather than for vengeance; and it "solves" the nuclear weapon threat once and for all. That the acquisition of defenses could be highly provocative—witness the American reaction to knowledge of the Soviet civil defense program, which is reviewed in chapter 5—appears to be unacceptable to supporters of the SDI.

These weapons could, in turn, be part of a more profound shift of the U.S.-Soviet military competition into space. Since U.S. reconnaissance, communication, and early-warning satellites play an important role in the U.S. strategic force posture, as do their counterparts in the Soviet Union, extending weapon systems to space, which seems certain to occur, could have profound implications for the way in which the strategic balance is assessed.

By the late 1970s experts were already envisioning orbiting particle-beam weapons that could disable early-warning satellites; defender satellites that could guard a missile-defense satellite cluster using laser weapons to destroy antisatellite missiles and space mines; and F-15 fighter aircraft that could launch antisatellite missiles at electronic intelligence-gathering (ferret) satellites.[24] In the case of the F-15, operational tests were begun in 1983.[25]

To be sure, the fundamental principles of war would still apply in space: measures and countermeasures and a premium on mobility, deception, and defense in an environment of high-accuracy weapon systems. How such a competition would affect both sides' ability to defend and control their forces is not clear, but both superpowers seem intent on finding out. The U.S. space shuttle program will be the basis for many military applications in space; the air force has formed a space command; and both the United States and the Soviet Union seem to be gearing up for a major expansion of their military space activities. Space is already seen not only as the high frontier but also as the last frontier, the key to the resolution of the military competition between the superpowers. By the early 1980s the Soviets had developed relatively primitive antisatellite capabilities against low-orbit targets but could not

24. See "The New Military Race in Space," *Business Week*, June 4, 1979, pp. 136–49.

25. William J. Broad, "Weapon against Satellites Ready for Test," *New York Times*, August 23, 1983.

yet demonstrate a capability against maneuverable or multiple targets or operate at the high, geosynchronous orbits of 20,000 miles used by American communication and early-warning satellites. It remains to be seen whether either superpower will conclude that it is in its political and military interests to constrain this competition, perhaps by restricting ASAT systems to low orbits, where their performance could be verified, or by signing a joint agreement declaring that satellite attacks are acts of war.

Finally, support is building in the defense community for a third generation of nuclear weapons.[26] The first generation included the early atomic and thermonuclear weapons of the 1940s and 1950s. The second generation consisted of the high yield-to-weight ratio warheads placed on Minuteman ICBMs and Polaris SLBMs in the 1960s. In the third generation of weapons, the blast, heat, and radiation effects of nuclear weapons would be controlled to limit the ensuing collateral damage. The enhanced radiation warhead, or neutron bomb, is cited as a prototypical third-generation device. Indeed, some believe that the combination of all three technological developments—directed-energy weapons used in space powered by nuclear detonators to produce minimum collateral damage—will revolutionize warfare and reinstitute defense as the primary military force. Moreover, the marriage of these technologies may point the way to a revised formulation of the so-called ASW breakthrough, which both governments have spent billions of dollars seeking to achieve. The preferred alternative to finding a submarine in the open ocean and destroying it may be to destroy the submarine-launched ballistic missile in its boost phase without first having had to locate the submarine.

Both governments are vigorously pursuing research on high-accuracy systems that could upset the present offense-dominated strategic balance. This activity is rooted in the insecurities of both societies, in organizational incentives, in the intrinsic appeal of the exotic technologies, and in the intense U.S.-Soviet political rivalry. Seeking unilateral military advantage as opposed to a stabilized military relationship surely dominates government policies in both Moscow and Washington. Thus, the likelihood of reciprocal unilateral restraint or negotiated limitations on these weapon programs seems remote. This reluctance to limit development will certainly be strong as long as one side, in this case the

26. See Judith Miller, "New Generation of Nuclear Arms with Controlled Effects Foreseen," *New York Times,* October 29, 1982.

THE BROOKINGS INSTITUTION WASHINGTON, D.C.
ISBN 0-8157-5964-9

United States, considers itself to hold a substantial technological edge. Should any of these systems develop into a credible BMD system, great diplomacy on the part of both governments will be needed to manage the transition to a new strategic condition. But for many years to come, these technologies will not fundamentally challenge the nuclear stalemate.

The heightened vulnerability brought about by new high-accuracy systems will, however, require the United States to rethink its declaratory policies and place greater demands than ever on its ability to maintain command and control of its nuclear forces in the event of war.

CHAPTER FIVE

U.S. Strategic Doctrine and Potential Response to War

We, we have chosen our path—
Path to a clear-purposed goal,
Path of advance!—but it leads
A long, steep journey, through sunk
Gorges, o'er mountains in snow.

Matthew Arnold, *Rugby Chapel*

WHEN PREPARING for war or predicting the likelihood of its occurrence, it is essential to assess both the capabilities and the intentions of the adversary. Some would argue that one should pay attention only to capabilities, since they at least are knowable, whereas intentions can change instantaneously. But clearly the two are interrelated. Sometimes a government formulates an intention and then seeks to acquire the capability to achieve it. At other times the acquisition of a new capability, perhaps with the help of an unanticipated technological advance, could give leaders new objectives that they had not previously entertained. Intentions, however, when formalized as doctrine, accepted by the public, and absorbed by those organizations responsible for exercising actions, usually become part of a state's capabilities. Intentions so integrated define the parameters of action and set limits on what one thinks one is capable of doing.

In the nuclear age, force levels (the kinds and quantities of weapons) and modernization (the rate at which new weapons are developed and produced and older weapons retired) are largely determined by military and political requirements. Bureaucratic interests, domestic political considerations, and the outcome of personal and organizational disputes within the defense policy community also exert powerful influence over what forces are actually deployed. But three factors are paramount in contemporary force planning: projections of the military capabilities of potential adversaries (discussed in the previous chapter), constraints placed on the use of nuclear weapons specified by arms control agreements (examined in the next chapter), and strategic doctrine.

A doctrine is a statement of fundamental government policy; a strategic doctrine is a statement of fundamental government policy about nuclear weapons.[1] Strategic doctrine defines the purposes of nuclear weapons, how and when they would be used (if at all), and the degree of confidence required in weapon effectiveness when planning for various contingencies.

The Soviet-American nuclear balance has evolved through the successive stages of American nuclear monopoly (1945–49), American superiority (1949–mid 1960s), and the present strategic stalemate. American strategic doctrine has undergone several shifts in emphasis and taken on a variety of labels during these periods. Since the 1960s, however, two objectives have dominated strategic doctrine: deterrence and stability. The goal of deterrence has been closely but not solely linked with assured destruction, which, as defined by former Secretary of Defense Robert S. McNamara, requires deploying a force able "to deter deliberate nuclear attack upon the United States and its allies by maintaining, continuously, a highly reliable ability to inflict an unacceptable degree of damage upon any single aggressor, or combination of aggressors, at any time during the course of a strategic nuclear exchange, even after absorbing a surprise first strike."[2] The goal of stability has meant the preservation of a strategic balance between the United States and the Soviet Union such that neither side would, at any time, perceive any advantage in initiating the use of nuclear weapons.[3]

Although members of the defense policy community have agreed on

1. It has already been noted that *strategic* is used in many different contexts. For purposes of this discussion the term is meant to convey the role of central strategic systems (such as ICBMs, SLBMs, and intercontinental bombers) and related systems in attacks on the U.S. and the Soviet homelands.

2. *Statement of Secretary of Defense Robert S. McNamara before the House Subcommittee on Department of Defense Appropriations on the Fiscal Year 1968–72 Defense Program and 1968 Defense Budget* (Government Printing Office, 1967), p. 38.

3. As noted in chapter 4, note 8, in the strategic literature a distinction is made between *crisis stability* and *arms race stability*. Crisis stability is concerned with force deployments that could provide one side or the other with an incentive to strike first in a crisis, either because it had a disarming capability or because it feared it was about to be struck itself and determined that its forces were in a use-it-or-lose-it situation. Arms race stability is associated with force deployments that could stimulate countermeasures by the other side. It can be argued persuasively that maintaining a position of perceived equality is a political objective of equal value to stability and assured destruction.

these objectives for many years, they have not agreed on how to obtain them. Thus, while American defense analysts have debated, often acrimoniously, the fine details of strategic doctrine, several inconsistencies have not received adequate attention. First, a sensible strategic doctrine designed in peacetime to deter a nuclear attack is not necessarily an accurate guide to what a government would do after a nuclear war had begun. Second, although one might like to think that U.S. policymakers first establish doctrine and then seek authorization from Congress to acquire the weapons to support that doctrine, in reality they usually promulgate doctrine to mesh with the existing force structure. Third, although the debates about doctrine revolve around deterrence and stability, nuclear weapons have so far been used solely as instruments of political coercion, for which the United States has no articulated guidelines. Fourth, the capacity to use high-accuracy weapons has led to changes in doctrine that have aroused fears in the public as deep as the comfort the new systems have given their supporters. The result is that strategic doctrine is far more an instrument for pursuing domestic and international political objectives than a precise plan for the use of nuclear weapons.

Deterrence and Doctrine before the Advent of Missiles

Although the concept of deterrence was not invented in the nuclear age,[4] it has been only since 1945 that national leaders have spoken of deterrence of an attack—a nuclear strike against the continental United States or nuclear or conventional aggression in Europe—as a principle of American policy.

In thinking about strategic deterrence, it is useful to review the familiar textbook definition of the term. We can say that A deters B if three conditions are met: A threatens to punish B if B pursues a particular course of action; B was planning to take this action before A's warning;

4. George Quester has described how national leaders employed the concept before the advent of nuclear weapons. See Quester, *Deterrence before Hiroshima: The Airpower Background of Modern Strategy* (John Wiley, 1966). The most comprehensive treatment of the changing formulation and implementation of the concept through the early 1970s is Alexander L. George and Richard Smoke, *Deterrence in American Foreign Policy: Theory and Practice* (Columbia University Press, 1974). The most helpful analytical dissection is William W. Kaufmann, *The Requirements of Deterrence* (Princeton, N.J.: Center of International Studies, 1954).

and B determines that A has the will and capability to carry out its threat and that the costs that B would incur from A's punishment would outweigh what B would gain from pursuing its original course of action.

The first condition is the communication of a threat. The communication may be explicit or implicit, sporadic or repetitive, but B must understand that A seeks to prevent B's action by threatening a counteraction. The second condition concerns actual intentions. The third condition concerns perceived capabilities—B's judgment about A's capability to carry out its threat.

Deterrence is a highly subjective, psychological phenomenon. Its success cannot be proven, since B's failure to act may not be tied to A's threat. Something definite can be said about deterrence only when it fails. Once it has failed, of course, it is a separate matter whether A should carry out its original threat. The purpose of that threat was to prevent B's action, not necessarily to punish B for taking the undesired action.

The first U.S. strategic doctrine after World War II that tied together objectives, resources, and constraints was the policy of containment enunciated most persuasively by George Kennan in his "long telegram" of 1946, published a year later in *Foreign Affairs*. But Kennan did not mention nuclear weapons in formulating strategy toward the Soviet Union. It was not until 1950 that the role of atomic weapons in U.S. national strategy was really addressed, in "NSC-68," a joint study that President Harry Truman requested of the State and Defense departments after the first Soviet nuclear explosion in 1949. The study called for a rapid increase in the stockpile of atomic weapons but did not discuss in any detail how, against what targets, or under what circumstances these weapons were to be used.[5] The study also called for a substantial buildup in conventional forces, strongly suggesting that atomic weapons were no substitute for nonnuclear forces for such purposes as deterring a Soviet conventional attack in Europe. Partly because the recommendations of NSC-68 were not made public for many years, the document is often relegated to the prehistory of U.S. strategic doctrine.

The history of U.S. strategic doctrine as we now know it began in 1954, after both sides had produced thermonuclear weapons, with Secretary of State John Foster Dulles's endorsement of massive retal-

5. The document was not declassified until 1975. See "NSC-68: A Report to the National Security Council," *Naval War College Review*, vol. 27 (May–June 1975), especially pp. 81–87.

iation as the essence of U.S. nuclear strategy. The United States adopted this posture to show off its technological strength and to provide a domestic political rationale for reducing defense manpower levels as well as the defense budget, objectives consistent with Republican party preferences at the time. Dulles noted in a speech at the Council on Foreign Relations in New York:

> Local defense will always be important. But there is no local defense which alone will contain the mighty landpower of the Communist world. Local defenses must be reinforced by the further deterrent of massive retaliatory power. . . . The way to deter aggression is for the free community to be willing and able to respond vigorously at places and with means of its own choosing.[6]

It was this declared policy that triggered an enormous public debate about the credibility of U.S. nuclear strategy. Dulles's remarks made clear that the United States was relying on nuclear weapons to deter and, if necessary, wage war against communist powers. They also implied that the United States was willing to strike population centers of the Soviet Union to retaliate against a first strike. This was not necessarily the case, judging from recently declassified material. As early as August 1950 and certainly by 1952, the Joint Chiefs of Staff had approved an emergency war plan for the Strategic Air Command (SAC) that had three basic objectives, noted by code names:

—BRAVO: the blunting of Soviet capability to deliver an atomic offensive against the United States and its allies;

—DELTA: the disruption of vital elements of the Soviet war-making capacity;

—ROMEO: the slowing of Soviet advances into western Eurasia.

There is little doubt that the United States would have used large numbers (600–750) of nuclear weapons to destroy the targets inherent in these objectives and that civilian casualties in the Soviet Union would have been in the tens of millions. But the principal targets were counterforce (military forces) and urban-industrial (factories), not civilian population centers. This "optimum mix" emergency war plan was accompanied by "several hundred strike plans," so the image Dulles had conveyed of wiping out Soviet society did not jibe with actual war plans. One significant aspect of these plans, however, was that they were totally under the control of the SAC commander, General Curtis LeMay,

6. See "The Evolution of Foreign Policy," *Department of State Bulletin*, vol. 30 (January 25, 1954), p. 108.

who alone could decide what targets to strike once he was given authority by the president.[7]

In the late 1950s the Eisenhower administration dropped the term *massive retaliation* from its political vocabulary and, in response to new ideas emanating from the air force and the Rand Corporation, refined war plans to concentrate more on military, industrial, and command and control targets. An important doctrinal document of the time, the Gaither Report, sent to President Eisenhower shortly after the Soviets launched Sputnik I in October 1957, called for major increases in defense spending, a nationwide civil defense program, increased air defenses, and an accelerated deployment of both land- and sea-based ballistic missiles. However, that report had no effect on U.S. strategic targeting plans.[8] By the time John Kennedy took office there were several thousand nuclear weapons deployed in Europe under control of the U.S. Army, and nuclear weapon use was no longer in the hands of the SAC commander alone.

The Shift to Flexible Targeting and Inflexible Rhetoric

Within two years of taking office Secretary of Defense Robert McNamara streamlined nuclear war planning further by establishing the nuclear weapons employment policy (NUWEP), to be issued by the secretary himself; recasting the single integrated operational plan (SIOP), developed initially in the Eisenhower administration, to specify the attack options to implement the NUWEP; and initiating the national strategic target list (NSTL), from which the SIOP was derived. Through the 1960s the target list was expanded and more options were included, although precise details of the NUWEP, SIOP, and NSTL are not known except to the small number of people with access to these highly classified documents. Yet the statements of the secretary, even in his classified posture statements, reflected a concern for first "damage limitation" and then "assured destruction," which masked the true capabilities of and plans for the strategic forces.

7. See David Alan Rosenberg, " 'A Smoking Radiating Ruin at the End of Two Hours': Documents on American Plans for Nuclear War with the Soviet Union, 1954–55," *International Security*, vol. 6 (Winter 1981–82), pp. 3–38, especially pp. 9, 25.

8. See U.S. Congress, Joint Committee on Defense Production, *Deterrence and Survival in the Nuclear Age (The "Gaither Report" of 1957)*, Committee Print, 94 Cong. 2 sess. (GPO, 1976).

In retrospect it was clear by 1965 that declaratory policy and actual guidelines for nuclear weapon use as translated into war plans conveyed very different impressions of where a U.S. nuclear strike would hit. The former stressed countervalue targets—population centers—and the latter stressed counterforce targets.

At the declaratory level, the policy of assured destruction depended on maintaining retaliatory forces that could survive the most severe initial attack and still destroy the attacker's armed forces, industry, and population. It was widely accepted that this capability must be made absolutely clear to potential adversaries. According to McNamara: "It is important to understand that assured destruction is the very essence of the whole deterrence concept. We must possess an actual assured destruction capability, and that capability also must be credible."[9]

To ensure the credibility of U.S. retaliatory forces, force planners followed two precepts. First, they made optimistic assumptions (from the attacker's perspective) about enemy effectiveness in a first strike and designed retaliatory capabilities based on these assumptions. Second, they designed independent retaliatory capabilities in the three separate offensive systems—land-based intercontinental ballistic missiles, submarine-launched ballistic missiles, and manned bombers—that compose the Triad. Because of their different modes of basing and penetration, the systems together provided synergism and protection not available from each system separately.

In the early years McNamara emphasized equally damage limitation and assured destruction. But he publicly abandoned the former and emphasized the latter as he witnessed the growth of Soviet forces, heard the demands—which he thought excessive—of the military for the weapons needed to destroy these forces, and realized the political costs of mounting a nationwide civil defense program. In other words, contrary to the popular view, assured destruction was related more to the size of strategic forces than to the doctrine for their use.

The assured destruction criterion did not go unchallenged, however. One school of criticism, emphasizing defense rather than deterrence, became prominent in the late 1960s during the debate about building the Safeguard antiballistic missile system. This school disputed the basic premise that the best way to prevent nuclear war is to make sure each side's population and industry is vulnerable to the other side's retaliatory forces. These critics advocated a policy based on the protection of U.S.

9. Robert S. McNamara, *The Essence of Security: Reflections in Office* (Harper and Row, 1968), p. 52.

population and industry through the massive deployment of antimissile and antiaircraft defenses, combined with heavy civil defenses. Supporters of this view contended that only by lessening the potential destruction of an opponent's attack could such an attack be deterred, whereas deterrence through an assured-destruction capability is unstable, uncertain, and the product of wishful thinking. This view failed to gain wide support because of the technical ineffectiveness and high costs associated with the building and operation of ABM systems; the realization that any such defenses could be offset with relative ease by increases in opposing offensive forces; and the high political, social, and economic costs of a major civil defense effort.

A second more potent criticism was that the doctrine of assured destruction guaranteed that any use of nuclear weapons would be virtual genocide and that this approach is inconsistent with American values and security interests. Critics argued that assured destruction was not a credible deterrent against limited attacks in which relatively few civilians would be killed, since to retaliate according to this criterion would result in national suicide. These critics proposed that emphasis instead be placed on developing forces that would permit the discriminate and flexible use of both nuclear and nonnuclear power as the most effective way to deter all kinds of threats. In 1974 these ideas were publicly endorsed as the notion of "flexible options" by Secretary of Defense James A. Schlesinger and were also supported by his successor, Donald H. Rumsfeld.

Although McNamara publicly continued to be associated with assured destruction—a massive, countervalue U.S. response to an initial Soviet attack—privately he never abandoned notions of damage limitation and urged the use of the SIOP for limited responses to limited provocations.[10]

10. McNamara's differing public and private views were reflected in other areas as well. He, of course, was also identified with the doctrine of flexible response, but this implied acquiring a range of military capabilities from counterinsurgency forces to strategic nuclear forces and was particularly associated with maintaining a credible deterrent posture for the European theater by advocating the escalation by NATO to first use of nuclear weapons if a Soviet conventional attack in Europe could not be contained with conventional forces. More recently McNamara has revealed that "in long private conversations with successive Presidents—Kennedy and Johnson—I recommended, without qualification, that they never initiate, under any circumstances, the use of nuclear weapons. I believe they accepted my recommendation." See Robert McNamara, "The Military Role of Nuclear Weapons," *Foreign Affairs*, vol. 62 (Fall 1983), p. 79. This matter is addressed in more detail in chapter 7.

Flexible Policies for Flexible Forces

The notion of flexible strategic options was thus not new when Schlesinger became secretary of defense. Schlesinger's enunciated policies, followed by the "countervailing strategy" of his successor, Harold Brown, were extensions of the basic ideas long embedded in actual war plans. The modifications of Schlesinger and Brown were incremental (for example, they called for more targets, fewer weapons per target, and more emphasis on command and control) rather than significant departures from existing practice.

In each of President Nixon's foreign policy messages to the Congress, he referred to the need for flexibility. In 1972 he stated: "No president should be left with only one strategic course of action, particularly that of ordering the mass destruction of enemy civilians and facilities. Given the range of possible political-military situations which could conceivably confront us . . . we must be able to respond at levels appropriate to the situation."[11] Secretary Schlesinger called for "a series of measured responses to aggression which bear some relation to the provocation, have prospects of terminating hostilities before general nuclear war breaks out, and leave some possibility for restoring deterrence."[12]

The doctrine of flexible options satisfied several interrelated objectives. The first of these was to enhance deterrence by being able to threaten a credible response to a variety of possible threats. The second objective was to achieve crisis stability by reducing a potential adversary's incentives to strike first in a crisis. This would be achieved by being able to respond to a limited strike with an attack in kind. The third objective was to control escalation—that is, limit damage and destruction in case deterrence failed and, in the process, decrease the likelihood that the initial use of nuclear weapons would escalate and result in civilian deaths or a full strategic exchange. This objective would apply to accidents and unauthorized acts as well as to deliberate limited attacks sanctioned by governments. The fourth objective was alliance cohesion—that is, to reinforce confidence in the American nuclear guarantee to European and Japanese allies by acquiring a variety of limited nuclear and nonnuclear response options. It was argued that European allies

11. Richard M. Nixon, *U.S. Foreign Policy for the 1970s: The Emerging Structure of Peace*, a report to the Congress (GPO, 1972), p. 158.

12. *Department of Defense Annual Report, Fiscal Year 1975*, p. 38.

would believe that the United States would be more likely to exercise these options than the policy of assured destruction in the event of a Warsaw Pact invasion of Western Europe.

Schlesinger and then Brown sought to improve the targeting flexibility of the strategic forces by expanding the number of preplanned options, by increasing the flexibility of command and control systems that permit missiles to be retargeted, by improving the accuracy of delivery vehicles, and by increasing the capability of strategic missiles to destroy hardened targets.[13]

Schlesinger and Brown and their colleagues did more than redefine deterrence and then translate this doctrine into force planning requirements. As McNamara had done before, they engaged in difficult organizational bargaining—within the executive branch and between the executive branch and the appropriate congressional committees—to arrive at force posture decisions, and then they formulated doctrines to generate public support for these decisions.

The acquisition of an effective counterforce capability and the difficulty of distinguishing between targeting flexibility and countersilo capability illustrate the problem of relating force posture to declaratory policy. As discussed in chapter 4, targets are categorized as either countervalue or counterforce. Countervalue targets include urban-industrial areas, which are commonly referred to as countercity targets, as well as nonurban civilian sites of economic, political, cultural, or historical significance. Counterforce, or countermilitary, targets include strategic nuclear forces as well as the other assets that comprise a nation's war-making capability, such as troop concentrations, airfields, logistics facilities, and transportation and communication systems. An important class of counterforce targets is hardened missile sites and command and control facilities, known as hard targets. The ability to destroy a large part of an opponent's land-based strategic forces is often referred to as a countersilo capability; the ability to destroy a large part of an opponent's entire strategic force is known as a disarming-first-

13. Schlesinger's declaratory policies are best defended in *Briefing on Counterforce Attacks*, Hearing before the Subcommittee on Arms Control, International Law and Organization of the Senate Committee on Foreign Relations, 93 Cong. 2 sess. (GPO, 1975). The articulation of Brown's policies may be found in Walter Slocombe, "The Countervailing Strategy," *International Security*, vol. 5 (Spring 1981), pp. 18–27. As Slocombe states: "It is *not* a new strategic doctrine; it is *not* a radical departure from U.S. strategic policy over the past decade or so. It *is*, rather, an evolutionary refinement and a recodification of the U.S. strategic policy from which it flows" (p. 24).

strike capability. From the outset, U.S. counterforce capability has been justified primarily as a way to maintain a stable strategic balance and not as a unilateral attempt to threaten Soviet strategic forces. It is within the context of this element of strategic doctrine—equivalence—that the evaluation of counterforce capabilities has taken place.

Maintaining a rough balance or equivalence in strategic forces has been a stated objective of American policy since the early 1970s, when Soviet strategic force levels matched and then began to exceed those of the United States. The United States' pursuit of an enhanced capability to destroy Soviet missile silos has generally been related to this objective. It is argued that if a substantial imbalance develops in countersilo capability or in other significant capabilities of the two strategic forces, the Soviet Union might wrest various concessions from the United States or its allies in spite of a secure American retaliatory capability. Proponents of equivalence believe the Cuban missile crisis was evidence of the importance of the balance of nuclear forces in a crisis and argue that the United States was able to force the Soviet Union to withdraw its missiles from Cuba because of the clear strategic superiority of the United States at that time. If the strategic balance is reversed in the future, it is argued, a similar retreat could be forced on the United States.[14]

The pursuit of an improved hard-target capability to deny the Soviet Union an advantage in this area of strategic power has been criticized on three grounds. First, it is argued that the Soviet Union would feel threatened by this capability, given its heavy investment in ICBMs, and would, if the U.S. capability was sufficiently developed, feel compelled to strike first or at least adopt a launch-under-attack posture for fear of being disarmed. The second criticism is that secure U.S. retaliatory forces and the great uncertainties and risk of escalation that would accompany any strategic exchange would prevent the Soviet Union from

14. This view has been articulated in Paul H. Nitze, "Assuring Strategic Stability in an Era of Detente," *Foreign Affairs,* vol. 54 (January 1976), pp. 215–16. However, as noted in chapter 3, note 11, several of the key participants in the crisis, including Robert McNamara, McGeorge Bundy, and Dean Rusk, believe that the conventional balance of forces alone determined the outcome of the crisis. What is unknowable, however, is whether the outcome of the crisis would have been different if the conventional force balance had been as it was while the nuclear arsenals of the two powers had been reversed. It seems plausible to argue that the United States *did* derive at least an implicit advantage because it enjoyed both nuclear and conventional superiority in the same theater of operation.

deriving any political or military benefit, regardless of its advantage in countersilo capability. There is thus no need for the United States to acquire a countersilo capability, since, strategically, it is valueless. Third—and this is a criticism leveled at the entire equivalence doctrine—the objective of balance actually permits the Soviet Union to dominate the design of U.S. strategic forces. According to this view, American force planners are rapidly becoming captive of a matching game in which every Soviet advance or projected advance must also be achieved by the United States, irrespective of its political or military utility or its cost.

Proponents of improved countersilo capabilities offer several rejoinders to these criticisms. They suggest that a secure Soviet SLBM force would preclude any Soviet sense of a need for preemption and thus would reduce any harmful effects of U.S. countersilo capabilities on crisis stability. Moreover, they argue that if the United States seriously threatens Soviet ICBMs, the Soviets would have added incentive to either transfer some of their ICBMs to sea or deploy less vulnerable mobile ICBMs. In either case, the threat to U.S. ICBM survivability would be correspondingly reduced and strategic stability enhanced. Finally, it is argued that high-accuracy ICBMs are inevitable because of technological developments.

A fourth criticism of acquiring a rapid, or "prompt" (that is, short flight time), countersilo capability in an ICBM force is more difficult to refute. If the forces are planned only for retaliation, it is not clear that a prompt response would be useful, since the attacker would presumably launch the remainder of his force on warning once alerted to the incipient retaliatory attack.[15] This argument has been countered primarily on the political and psychological grounds that in the minds of Americans and their allies roughly equivalent countersilo capabilities strengthen the overall deterrent posture of the United States in ways that an unbalanced situation in peacetime would not.

Given the many ways of evaluating the capability of strategic forces and the uncertainty about how they would perform in a nuclear war, equivalence as the criterion for force sizing (that is, the number of forces to be deployed) is an ambiguous concept. Ultimately, this concept is

15. The logic of the various uses of prompt countersilo retaliatory capabilities is spelled out in Albert Carnesale and Charles Glaser, "ICBM Vulnerability: The Cures Are Worse Than the Disease," *International Security*, vol. 7 (Summer 1982), pp. 80–82.

subjective rather than objective. Also, given the intense politics of the defense budgeting process, debates among specialists will inevitably place doubt in the public's mind about whether so-called equivalence has been achieved.

A question now, in the 1980s, is whether the Reagan administration is trying to regain nuclear superiority over the Soviet Union. This question is not terribly interesting in a military sense as long as existing technological assumptions remain valid. Even if the United States were to deploy the MX missile, the Trident II SLBM, the B-1, and then the Stealth bomber and there were few Soviet deployments in kind (except to reduce ICBM vulnerability), little in fact would change. There would still be a nuclear stalemate because of the multiplicity of invulnerable retaliatory forces. By this logic, the likelihood of nuclear war would be unaffected. Such deployments would not seriously affect crisis stability, although they would stimulate the arms competition.

Successive administrations have interpreted and reinterpreted American strategic doctrine to satisfy a variety of domestic and international political objectives and force-sizing objectives that have not often been closely related to targeting policy. Moreover, it is not at all certain that in the event of nuclear war, American leaders who survive the initial attack would want to limit themselves to preplanned options. They would more likely make ad hoc decisions, which is normal in a crisis, and explore a wide range of options. To be sure, preconceived policies would limit the range of options but would not necessarily define all of them.

Given the highly tenuous relationship between American strategic doctrine and probable American behavior in a nuclear war, there is reason to be cautious in evaluating Soviet strategic doctrine. First of all, sources are not plentiful, and those that are available are difficult to interpret.[16] When a Soviet document is published, it is hard for experts

16. Two dated but informative documents are V. D. Sokolovskiy, *Soviet Military Strategy*, 3d ed. (Crane, Russak, 1975), especially pp. 5–46, 172–211; and A. A. Sidorenko, *The Offensive* (GPO, 1970), especially pp. 5–70, 109–37. A recent review article that evaluates most of the Soviet writings available to Western analysts is Dan L. Strode and Rebecca V. Strode, "Diplomacy and Defense in Soviet National Security Policy," *International Security*, vol. 8 (Fall 1983), pp. 91–116. The authors conclude that "there appears to be an element of real conflict in the formulation of military doctrine. Moreover, differences in the public treatment of military doctrine reflect broader disagreements over national security policy between those who stress the utility of diplomatic initiatives in influencing the policy of rival powers and those who stress the utility of unilateral applications of Soviet strength" (p. 115). Because of the inherent uncertainties in evaluating Soviet military writings, they are not examined here in detail.

on Soviet military doctrine to agree on its implications, even when it accurately reflects professional military intentions and capabilities. Statements of Soviet civilian leaders confirm all U.S. interpretations, from the most to the least malevolent.

The Soviets take their military missions seriously and are seeking to deploy forces that would be most useful in a time of war. Their doctrine seems to emphasize the capacity to preempt should leaders believe that an attack against the Soviet Union is imminent. The Soviets prefer, as they have historically, quantitative superiority where feasible in all measures of military capability. Soviet doctrine supports capabilities to hit counterforce and command, control, communication, and intelligence (C^3I) targets and stresses a combined arms approach involving nuclear, conventional, and chemical weapons to achieve military objectives. Moreover, great doctrinal stress is placed on development of both active and passive defense systems.

These are the ideas of the professional military who are responsible for waging war in the nuclear age. It is a separate and largely unknowable matter whether Soviet political leaders believe a nuclear war can be won or whose views would matter the most once a decision to use nuclear weapons had been made. The maintenance of secure U.S. retaliatory forces would in all probability prevent the world from ever finding out.

Political Uses of Nuclear Weapons

The U.S. government has used nuclear weapons to achieve political goals for which it has had no doctrinal guidance. Current debates about an American "window of vulnerability" suggest that some citizens are less concerned about the ability of the United States to flex its nuclear muscles than they are about the possibility that the Soviet Union will also use nuclear weapons for political purposes. The related concepts of security and coercion can help clarify why the notion of a window of vulnerability seems to be baseless.

Security is the freedom from fear and from harm. It is also the freedom to act, to pursue objectives without fear of being thwarted by opposing forces. Now, of course, no state is totally secure; security is a relative concept. Any state's freedom of action is necessarily limited by its resources and capabilities and by the ability of other states or groups of states to block a given course of action.

Coercion, on the other hand, is the ability to restrain or dominate. It is the use of force or threats to overpower an individual will or to compel that individual to behave in a particular way. Security, then, is in part the freedom from coercion.

So-called nuclear diplomacy has successfully influenced the outcome of a crisis in instances when the United States had clear strategic and local-force superiority. Given the present state of the strategic balance, it would be dangerous for American leaders who believed in a window of vulnerability to permit themselves to be coerced by Soviet officials who threatened to attack the U.S. ICBM force. Such a threat is not credible because of the ability of the United States to launch under attack, the uncertain success of a Soviet attack, and the retaliatory forces that U.S. leaders would still have at their disposal. It is indeed hard to conceive of even a nonnuclear action taken or a threat made by the Soviets that would be more credible because of the vulnerability of U.S. ICBMs. The window of vulnerability, therefore, seems to be illusory.

In several instances when the United States held nuclear superiority over the Soviet Union—during the Korean War and the Quemoy-Matsu crisis in the 1950s and the Berlin and Cuban missile crises in the 1960s, for example—the United States used nuclear weapons coercively to get its way.[17] Such nuclear diplomacy was carried out through either an explicit or implicit threat of nuclear weapon use, a change in the alert status of U.S. forces, or the redeployment of nuclear forces. As the nuclear balance has shifted and American superiority has slipped, the United States has resorted to these actions much less frequently. The raising of the U.S. military alert status to defense condition 3 during the October 1973 war in the Middle East was the last publicized American act of nuclear diplomacy in a crisis. In this instance, the United States responded to what President Nixon and Secretary of State Kissinger judged to be a Soviet threat to intervene on the ground in the Sinai with some American nuclear saber rattling. It is still unclear how credible either the Soviet threat to intervene or the American response actually was. But from Kissinger's personal account, American domestic political weakness (that is, the Watergate affair) played a role in the American response: "There was some desultory talk about whether the Soviets would have taken on a 'functioning' President. I said: 'We are at a point

17. See Barry M. Blechman and Stephen S. Kaplan, *Force without War: U.S. Armed Forces as a Political Instrument* (Brookings Institution, 1978), pp. 47–49. A total of nineteen incidents have been recorded.

of maximum weakness but if we knuckle under now we are in real trouble.' "[18] What is most striking in the accounts of this and other episodes of nuclear diplomacy is how vague the motivations of U.S. policymakers have been. They do not seem to have clearly thought through all the consequences of their actions or how they would respond if the Soviet Union were not successfully coerced.

The Soviets have threatened to use nuclear weapons on fewer occasions than the Americans have. These threats have ranged from their pledge during the 1958 Taiwan Straits crisis to supply the People's Republic of China with nuclear weapons if China were attacked by the United States to references to nuclear weapon use during the 1959 and 1961 Berlin crises.[19] Some officials in Washington judged Soviet officials' reference to the Soviet-Iranian Friendship Treaty during the 1978 Iranian crisis as a Soviet exercise in coercive diplomacy (although nuclear weapons were not mentioned in the Soviet statement) to deter the United States from intervening militarily on behalf of the shah.

The superpowers, with their present deployments of large and sophisticated nuclear forces, have reached a nuclear standoff. If both countries maintain this state—and they probably will, barring a major technological breakthrough, given the domestic forces that influence nuclear weapon policy in the two countries—each side will thus be denied the ability to exercise nuclear coercion against the other.[20] Both

18. Henry Kissinger, *Years of Upheaval* (Little, Brown, 1982), p. 589.

19. See David Holloway, *The Soviet Union and the Arms Race* (Yale University Press, 1983), pp. 81–108.

20. This judgment, of course, cannot be verified. The delicacy of the nuclear balance at the central strategic level has always been at issue, even among the pioneers of American strategic thought. At the regional level, the evidence of the last decade supports this judgment as do the views of two of the leading contributors to the literature on limited war, neither known for their softheadedness. Robert Osgood has stated that "outside Europe the credibility (in American eyes) of initiating the use of nuclear weapons under any circumstances seemed to reach its high point in 1954, during the Quemoy and Matsu crisis and the fall of Dienbienphu, and has steadily declined ever since." See "The Post War Strategy of Limited War: Before, During and After Vietnam," in Laurence Martin, ed., *Strategic Thought in the Nuclear Age* (Johns Hopkins University Press, 1979), p. 104. William Kaufmann has stated: "Even where both sides have large numbers of nuclear weapons (that is, thousands), it remains possible—at least on paper—for one side to achieve an exploitable military superiority over the other, although this kind of superiority is much more likely to be achieved at the strategic than at the tactical level. Despite this possibility, any nuclear exchange, strategic or tactical, would be fraught with the danger of escalation to unrestricted bombing and unprecedented civilian and military damage. . . . The idea that the United States, or for that matter the Soviet Union, is straining at the nuclear leash or is likely to launch nuclear weapons

sides will also continue to be careful that their armed forces do not engage each other in combat, for fear that such engagement would or could lead to a nuclear exchange.

Still, most of the defense community is nervous about how the Soviets might try to exploit their temporary advantage in countersilo capability. Some analysts expect that such exploitation would be political rather than military and that the United States is right to acquire forces that could deter a Soviet attack against a variety of U.S. targets. However, those in favor of arms control or a nuclear freeze disagree with this view.

These people, and those who see nuclear weapons as essentially a moral issue, believe that increased selectivity and flexibility lower rather than raise the nuclear threshold. Acquiring more nuclear weapons simply makes nuclear war more thinkable and thus more likely. The prevailing view in this community is that weapons such as the MX missile, which can destroy hard targets promptly, are especially pernicious. Such weapons, it is argued, are only useful as first-strike weapons. If the United States was struck first, it would have no need for the prompt responsiveness of these weapons because the attackers would launch a third strike on warning and the second-strike forces would destroy only empty silos. Those who hold this view do not consider destroying the capability of Soviet silos to reload missiles after the initial firing as significant. Finally, these people believe that the U.S. and Soviet joint accumulation of roughly 60,000 nuclear weapons has reached a level completely beyond any rational political or military purpose. Putting aside the rationale for specific systems, there is a general sense that "enough is enough" and that disarmament should now begin.

The rhetoric of the Reagan administration, especially between 1981 and 1983, produced public concern that the likelihood of nuclear war was increasing. Peace groups in Europe, nuclear freeze movements in the United States, and a variety of groups "for social responsibility" were formed to call attention to the U.S.-Soviet nuclear arms buildup. Students from grade school to graduate school expressed concern with

---

except in almost unimaginable circumstances is simply at variance with these realities. The decision is too hard, as anyone who has come at all close to it in the last twenty years knows only too well. Talk about an early use of nuclear weapons—even when it appears in supposedly serious Pentagon studies—is either the equivalent of cocktail party conversation or sheer bluff." "Defense Policy," in Joseph A. Pechman, ed., *Setting National Priorities: Agenda for the 1980s* (Brookings Institution, 1980), p. 294. Kaufmann served as a principal consultant to the secretary of defense from 1961 through 1980.

these issues and, for the most part, did not seem to believe in the need to modernize U.S. strategic forces. That the National Conference of Catholic Bishops criticized on moral grounds the underlying premises of the Reagan policy was significant in itself, since this policy was supposedly formulated in part as a moral alternative to assured destruction.

The defense community has paid scant attention to the views on nuclear issues of various nonspecialist groups, and for good reason. With the notable exception of protests over atmospheric testing that led to the 1963 Limited Test Ban Treaty and the opposition to ABM system deployments in New England in the late 1960s, the public has until now been conspicuously uninvolved in the American nuclear weapons debate. In the early 1970s the absence of published articles on nuclear-related issues could have been evidence that the public was "forgetting about the unthinkable."[21] But now the subject is debated routinely in all sorts of educational and community forums. George Quester, a well-published student of nuclear weapons policy, used to say somewhat whimsically that American policy would be in trouble when his grandmother inquired about the circular error probability of an SS-9 Soviet ICBM. We are not far from this condition today. By mobilizing themselves politically, concerned citizens could generate enough congressional support to thwart procurement of weapons in the strategic modernization program and induce the administration to adopt positions more conducive to arms control. It should be emphasized, however, that from 1945 through 1984 public opposition was never the principal cause of the termination of any nuclear weapon program.

In general, defense experts see the Soviet Union's prompt countersilo capability as the most dangerous development of recent times. Concerned citizens are more disturbed by the bilateral accumulation of nuclear weapons and by the perceived belligerence of the American government, reflected both in declaratory policies and in weapon deployment decisions. Therefore, a fundamental tension exists between trends in strategic doctrine and general public perceptions. While defense professionals emphasize more carefully controlled and limited use of nuclear weapons, either motivated by their own judgments about the credibility of deterrence or out of the need to justify new weapon systems, much of the public interprets these statements as a dangerous

21. Rob Paarlberg, "Forgetting about the Unthinkable," *Foreign Policy*, no. 10 (Spring 1973), pp. 132–40.

drift toward nuclear war. Citizens do not wish to be reminded of nuclear weapons and the prospect of their use. It is probably not an exaggeration to conclude that official U.S. statements about nuclear weapon use have done more than any Soviet act to stimulate peace and antinuclear groups in Europe and the nuclear freeze movement in the United States.

Several years ago two colleagues and I noted:

> Exercising prudence in the discussion of nuclear weapons policies is one relatively simple unilateral action which can contribute substantially to the goals of arms control. . . . It is crucial that the use of nuclear weapons be inhibited and, in the event of their use, that further use be halted as rapidly as possible. To discourage initial use, the firebreak between conventional and nuclear weapons must be preserved and indeed reinforced. This can be achieved in part by adopting a declaratory policy in which the possibility of using nuclear weapons is raised only by the President and is treated invariably with the awe it deserves. A balance should be struck in which allies are reassured of the strength of American security guaranties, including the use of nuclear weapons, without conveying in a cavalier fashion the notion that limited war is a readily feasible option.[22]

The intensity of public concern about nuclear weapons seems to confirm that this indeed is sensible advice.

The Possibility of War

None of the open debate about nuclear weapons addresses the question of our capacity to act in the event of a nuclear war. To answer this it is pertinent to review what happened when the United States dropped an atomic bomb on Japan in 1945, explore how a similar event might happen again, and address the questions of command and control and civil defense.

Hiroshima and Nagasaki

In the summer of 1945 President Truman did not know how much longer Japan would stay in the war. After Tokyo failed to reply to a joint ultimatum of the United States, the United Kingdom, and China in late July 1945, Truman approved the use of the atomic bomb after August 3 if Japan had not surrendered by then. As Truman later recalled:

22. Paul Doty, Albert Carnesale, and Michael Nacht, "The Race to Control Nuclear Arms," *Foreign Affairs*, vol. 55 (October 1976), p. 130.

Let there be no mistake about it. I regarded the bomb as a military weapon and never had any doubt that it should be used. . . . In deciding to use this bomb I wanted to make sure that it would be used as a weapon of war in the manner prescribed by the laws of war. That meant that I wanted it dropped on a military target. I had told [Secretary of War Henry L.] Stimson that the bomb should be dropped as nearly as possible upon a war production center of prime military importance.[23]

Stimson's staff produced a target list and after Kyoto was ruled out because of its cultural and religious significance, the order of priority became Hiroshima, Kokura, Niigata, and Nagasaki. At 8:15 a.m. on August 6, 1945, the B-29 *Enola Gay* released an atomic bomb at an altitude of about 32,000 feet, and the bomb exploded forty-three seconds later at 2,000 feet. The bomb, nicknamed Little Boy, used the uranium isotope U-235 in a gun-type assembly *never previously tested*. The bomb weighed four tons, released energy equivalent to thirteen kilotons of the high explosive trinitrotoluene (TNT), and immediately destroyed about two-thirds of the city. Three days later at 11:02 a.m. a second bomb was dropped on Nagasaki after the original target, Kokura, was passed over because of heavy cloud cover. This bomb, nicknamed Fat Man, used plutonium-239 in an implosion assembly identical in design to a device tested successfully at Alamogordo, New Mexico, on July 16. This bomb weighed four-and-a-half tons, had an explosive yield equivalent to twenty-two kilotons of TNT, and destroyed about half the city.

Physically, the atomic bomb produced extraordinarily high temperatures—several million degrees at the point of burst. In such an environment all materials became an ionized gas, electromagnetic waves of very short lengths were released, and a fireball formed. The fireball in turn emitted thermal radiation—about 35 percent of the total energy produced in the explosion. The intense heat of the fireball not only burned everything in its path but also produced enormous air pressure, which created powerful shock or blast waves.

The thermal radiation created surface temperatures of 3,000 to 4,000 degrees Centigrade for a distance of three-and-a-half to four kilometers from the center of the blast (called the *hypocenter*). Light-colored objects, which reflect heat and light, received considerably less intensive thermal burns than dark-colored objects, which absorb heat and light. Blast waves damaged and destroyed structures by virtue of the outside

23. Harry S. Truman, *Memoirs: Year of Decisions*, vol. 1 (Doubleday, 1955), pp. 419–20.

pressures created by the strong winds that preceded the shock waves and by vibrations of the structures themselves. The heat, blasts, and fires together destroyed more than each of these would have destroyed separately. In particular, both Hiroshima and Nagasaki experienced firestorms, violent whirlwinds that caused fires to spread farther and for longer periods than any ordinary fire.

The degree of damage decreased with distance from the center of the blast. In Hiroshima, 92 percent of more than 75,000 structures were damaged to some extent, and the greatest amount of destruction was within two kilometers of the hypocenter. The conclusion was that "the whole city was ruined instantaneously."[24]

The people of Hiroshima and Nagasaki experienced immediate and long-term consequences. At first they suffered thermal burns, bodily trauma, and radiation illness. Everyone who received severe thermal injuries from being without protection within one kilometer of the hypocenter died within a week. Of those who were between one-and-a-half to two kilometers from the hypocenter, 14 percent of the shielded and 83 percent of unshielded individuals died immediately.[25] Bodily trauma was caused directly by the blast and by falling buildings.

Radiation illness is caused by large doses (greater than ten roentgen[26]) of beta and gamma rays and neutrons released by the uranium and plutonium in a nuclear explosion. The radiation causes cell destruction and molecular alteration. The initial symptoms of radiation illness in Hiroshima and Nagasaki included nausea, vomiting, abnormal thirst, anorexia, general malaise, high fever, and diarrhea. These symptoms were most prevalent during the first week after the explosion. From the second to the fifth week people's hair began to fall out. Because the radiation damages blood cells, particularly in bone marrow, hemorrhages were also common.

Months after the attacks, disturbances were noted in the reproductive functions of both men and women, and pregnancy disorders materialized.

24. The Committee for the Compilation of Materials on Damage Caused by the Atomic Bombs in Hiroshima and Nagasaki, *Hiroshima and Nagasaki: The Physical, Medical, and Social Effects of the Atomic Bombings* (Basic Books, 1981), p. 61. (Hereafter *HN*.)

25. *HN*, p. 120.

26. A roentgen is the official unit of measurement for radiation. It is defined as the amount of radiation that produces in one cubic centimeter of dry air under standard conditions of temperature and pressure one electrostatic unit of charge.

Gonads and endocrine glands were acutely damaged. Malfunctioning of the central nervous system was not uncommon.

Years after the explosions, hematological and cardiovascular disorders, cancer (especially leukemia), liver dysfunctions, endocrinological diseases, and adverse genetic effects in children and even grandchildren of the victims are showing up.[27] The realization that the nuclear explosions of 1945 could have a significant impact on the health of people born decades later is certainly one of the most telling statements about the unique destructive qualities of nuclear weapons.

The bombs dropped on Hiroshima and Nagasaki also caused major social and psychological disruptions. The breakdown of community and family was virtually complete. The following passage graphically describes the situation in Hiroshima:

> Prefectorial office, city hall, fire departments, police stations, national railroad stations, post offices, telegram and telephone offices, broadcasting station, and schools—were totally demolished and burned. Streetcars, roads, and electricity, gas, water, and sewage facilities were ruined beyond use. Eighteen emergency hospitals and thirty-two first aid clinics were destroyed; and most of the personnel needed to restore them to use were killed or injured. As approximately 90 percent of all medical personnel in the city were dead or disabled, and the hospitals were destroyed or damaged, all medical care was thrown into confusion.[28]

Acquiring food to eat and water to drink was a major task, and survivors who found food and drink were unaware of the dangers of consuming irradiated foodstuffs.

It took several years to rebuild life in these two cities; the rebuilding was only possible because of the medical and physical support received from outside sources. There is no way to estimate how many years, or decades, it would have taken the survivors to rebuild their city if they had not got help. This highlights the difficulties of recovering from a massive multicity nuclear attack in which virtually all survivors would have to rely solely on themselves.

Surrounded by death and destruction, with many of the living carrying grotesque scars and deformations, and with the collapse of community and family, A-bomb victims lost their sense of psychological equilibrium. They felt apathetic toward life and in many cases would have rather died than continued to live. Confusion, emptiness, panic, uncertainty, and

27. See *HN*, pp. 186–332, for a detailed summary of the medical evidence marshaled to date.

28. *HN*, p. 379.

indifference toward even basic hygiene lasted for months among thousands of victims.

In sum, the statistics of 140,000 deaths in Hiroshima and 75,000 deaths in Nagasaki in no way captures the enormity of destruction nuclear weapons caused for these cities and their people.

Projections about Nuclear War

If nuclear weapons should be used in a Soviet-American war, how might the war start and how might the weapons be used? Four scenarios seem plausible. Nuclear war might come as a bolt from the blue or it might be an escalation of a conventional war, a preemptive strike in a deep crisis, or an accident. For the first scenario to happen Soviet leaders would have to believe that a disarming first strike could be successful. But without a breakthrough in ASW technology this is not possible now. Despite the enormous sums the Soviet Union has spent so far, no combination of passive and active detection systems poses a serious threat yet to the U.S. SSBN fleet.

A bolt-from-the-blue attack might also be possible if either side had a leakproof defense. Here again the technology is simply not at hand. In this era of MIRVs, saturation attacks, decoys and electronic countermeasures, and the vulnerability of BMD sensing devices make it unlikely that a workable defense could be deployed to protect population centers or even hardened military targets from a sophisticated attack. (However, ballistic-missile defenses with certain ICBM deployment patterns such as multiple protective shelters would increase costs for the attacker.) New technologies would have to be mastered and deployed before defense could be more powerful than offense.

In a third and more restrictive bolt-from-the-blue scenario, popularized by Paul Nitze, the Soviets would launch an attack on U.S. ICBMs, SAC aircraft, and SSBNs. A successful attack would still leave the United States with a residual force of perhaps 3,000 warheads based on the SSBNs at sea and on a few surviving ICBMs and long-range bombers. Nitze has argued that in such a situation the U.S. national command authorities would avoid retaliatory attacks for fear of Soviet reprisals in a third strike. The absence of a credible countersilo retaliatory capability would leave the command authorities with no adequate response.

This logic is questionable on several grounds, however. First, a U.S. retaliatory attack on Soviet silos would probably be of limited value since the Soviets would be expecting to launch under attack, having themselves already initiated nuclear war. Second, the United States could strike several military and industrial targets in retaliation, a move that would be militarily effective and show American resolve to proceed up the ladder of nuclear escalation. Third, it seems implausible that the highly conservative and risk-averse Soviet leadership would gamble that the United States would not retaliate.

A variation on the bolt-from-the-blue theme concerns an initial Soviet strike on the U.S. command and control system that would leave command authorities uncertain about the nature of the attack. Although this approach, if implemented successfully, would complicate U.S. decisionmaking, it would also be a formidable strategic warning to Washington and would leave the fate of Soviet society in the hands of the enemy. Moscow must see this as an extremely risky strategy with an uncertain payoff. A bolt-from-the-blue attack, even if restricted to counterforce targets, would still call for the detonation of several thousand warheads to make any military sense. This would almost certainly lead to countercity strikes and an all-out nuclear exchange. This scenario, while possible, is highly improbable.

The use of nuclear weapons in the escalation of a conventional war raises somewhat different prospects. A conventional war in Europe started by a Warsaw Pact attack against NATO forces could result in the use of nuclear weapons in at least three ways. NATO could, as former Secretary of State Alexander Haig suggested, launch a limited and highly discriminating attack against a single valuable Pact target to demonstrate resolve early in a conflict and persuade Moscow to stop hostilities before full-scale nuclear war ensues. In a more advanced stage, if NATO forces were clearly being defeated, nuclear forces could be used to interdict the attacking Pact armies and to strike second-echelon forces and other important targets in Eastern Europe. (Admittedly, though, various war exercises indicate that NATO would not gain from such an action.) Finally, the Soviet Union could use SS-20s and other prompt counterforce weapons either preventively or preemptively to disarm NATO of its nuclear escalatory capabilities and to ensure that the Soviets retain control of escalation throughout a conflict.

It is conceivable that a U.S.-Soviet conventional war begun outside Europe could result in some casualties or in a prolonged nonnuclear war. An example would be a case in which Israel attacks Syrian surface-to-air missile sites and kills Soviet advisers, the Soviets respond by striking at Israeli air forces, and American and Soviet forces come to blows protecting their allies' forces. To the extent that American and Soviet leaders sought to continue the conflict, they would probably do so at the conventional level rather than escalate to the use of nuclear weapons in the initial area of conflict, since a conventional war is thought to be more controllable than a nuclear war.

A preemptive strike in a crisis also raises serious obstacles for the attacker. A crisis usually implies a pervasive sense that an important point has been reached and that the path subsequently chosen will significantly affect the future; a departure from standard operating procedures and reliance instead on ad hoc decisionmaking processes; the use of experts to help resolve the problem; provision of face-saving options so the adversary can retreat rather than act out of desperation; and great urgency. A crisis also generally implies the understanding that it cannot merely be an exercise in avoiding disaster but is an opportunity to realize gains or take initiatives not feasible under normal circumstances. If an intense U.S.-Soviet crisis develops anywhere in the world, leaders would most likely try hard to keep channels of communication open to reduce the chance of misunderstanding and convey both capabilities and intentions in a way that would defuse the crisis so that war could be avoided.

Accidental war, a matter of great concern in the 1950s and early 1960s, is no longer a fear within the strategic community since permissive action links and other means of enhanced control over nuclear weapon use have been introduced. Nonetheless, as indicated by the relatively frequent failures of the North American Air Defense Command (NORAD) early warning system, the launching of nuclear forces either because of a system malfunction or by unauthorized personnel cannot be disregarded. Under such unfortunate circumstances the "attacker" would need the ability to communicate to the adversary the nature of the malfunction and the capacity to recall, disarm, or destroy the delivery vehicle before it reaches its target. The other side would need to be able to respond to an accidental nuclear attack in a highly circumscribed fashion, in the event the attack could not be voided.

Command, Control, Communication, and Intelligence

If nuclear war comes, U.S. leaders will be preoccupied with three tasks: stopping the war quickly without catastrophic military and political defeat; ensuring that if U.S. nuclear forces are used, they are used precisely as the leaders wish them to be used; and protecting citizens to whatever extent is possible.

If a nuclear war is to be ended, U.S. leaders must have someone who can communicate with the adversary, and they must ensure that an adversarial leadership survives with whom they can communicate. It is worth noting the lessons Robert Kennedy claimed to have learned from the Cuban missile crisis of October 1962:

—the need for time to debate with fellow officials and to argue through the options available;

—the need for multiple advocacy so that the president is not captive to a single point of view;

—the need for support of allies; and

—the importance of placing one's self in the adversary's shoes, of not humiliating him, and of leaving him a way out of the crisis short of disaster for all parties.[29]

Moreover, it would be crucially important to build the adversary's confidence in one's willingness to exercise restraint.[30] If the adversary believes his enemy has lost all sense of restraint, he is likely to act unrestrained as well. In the context of a nuclear war, this would be cataclysmic.

To meet these requirements as well as to have the ability to use nuclear forces in particular ways (for example, in a less-than-escalatory nuclear response to a first strike in order to show the adversary resolve plus a willingness to deescalate), the command structure of the United States must itself be able to survive an initial nuclear attack. Yet for many years the American strategic debate largely ignored this problem and instead focused on the capabilities of particular weapon systems, the doctrines for their use, and the means by which they could be controlled. Important decisions were made regarding force posture, and declaratory policies were enunciated. Implicit in all this discussion and decision was one crucial assumption: the systems would function as they were intended to. The proper commands would be conveyed; the responses would be appropriately controlled. There is a vast technical community that is responsible for ensuring that the command, control,

29. Robert F. Kennedy, *Thirteen Days: A Memoir of the Cuban Missile Crisis* (Norton, 1969), pp. 111–28.

30. A stimulating paper on how to approach this problem is Thomas C. Schelling, "Confidence in Crisis," *International Security*, vol. 8 (Spring 1984), pp. 55–66.

communication, and intelligence network for U.S. strategic forces does in fact operate according to specified criteria. Much effort has been expended and large sums of money have been spent on these tasks. But these are technical and procedural matters that have rarely engaged the policy community.

Only since the late 1970s have nongovernment defense specialists begun paying attention to C³I policy problems. Before the late 1970s, few analysts were aware of the characteristics of the C³I system and its strengths and weaknesses, and fewer still wrote about them.[31] Of greatest significance both inside and outside the government has been the shift from trying to ensure that the system would function properly *before* a nuclear attack so that *all* U.S. retaliatory forces could be activated to trying to ensure continued operation *during* a nuclear attack so that only *selected* nuclear forces would be used. The latter is a much more demanding task.

The C³I system consists of a complex network of satellite sensors, including ground, ship, and airborne radars and command posts. A vast array of communication links and digital computers transmit the information received to appropriate authorities organized under the worldwide military command and control system. The WWMCCS in turn provides essential information to the nation's civilian and military leaders, the national command authorities (NCA). Consequently, for U.S. military forces to be properly used, it is critical that the C³I system serve the needs of the command authorities and their subordinate commands.

These needs are summarized here.

Early warning of a nuclear attack. This function is performed primarily by monitoring Soviet and Chinese ICBM launch sites and the patrolling

31. John D. Steinbruner identified at an early stage the emerging technical weaknesses of the system and their highly significant policy implications. See "National Security and the Concept of Strategic Stability," *Journal of Conflict Resolution*, vol. 22 (September 1978), pp. 411–28, and, more recently, "Nuclear Decapitation," *Foreign Policy*, no. 45 (Winter 1981–82), pp. 16–28. Beyond the conceptual problems in maintaining an effective C³I system, the full range of operational questions has now been treated comprehensively in Desmond Ball, *Can Nuclear War Be Controlled?* Adelphi Paper 169 (London: International Institute for Strategic Studies, 1981), and Carnegie Panel on U.S. Security and the Future of Arms Control, *Challenges for U.S. National Security: Nuclear Strategy Issues of the 1980s*, A Third Report (Washington, D.C.: Carnegie Endowment for International Peace, 1982), pp. 85–134. See also *Department of Defense Annual Report, Fiscal Year 1983*, pp. III-77 to III-89, and Paul Bracken, *The Command and Control of Nuclear Forces* (Yale University Press, 1983). These references deal primarily with command, control, and communication (C³) and less with the intelligence function. Similarly, the discussion that follows emphasizes C³ rather than C³I.

areas of Soviet SSBNs to detect heat from engine ignitions and by tracking stations that identify missiles in the early stages of their ballistic trajectories. Since ICBM flight times from the Soviet Union to the United States are less than thirty minutes and depressed-trajectory SLBMs could reach American coastal cities in perhaps five minutes, the system needs to provide the NCA with clear information in a few minutes.

Attack assessment. The U.S. response to a nuclear attack depends on both the assessment of the attacking force before weapon detonation and an accurate evaluation of the magnitude and effectiveness of the attack once it has been carried out. Currently, both pre- and post-attack assessment capabilities are highly imprecise.

NCA survival. It does no good to transmit early warning and attack assessment information to decisionmakers if the NCA itself has been destroyed. This problem of protecting the president, the secretary of defense, and the chairman of the Joint Chiefs of Staff, or their successors, has been addressed by the establishment of several alternate land-based and air-based command posts. The ability of the NCA to both survive and control the forces at their command is one of the major vulnerabilities of the present system.[32]

Transmission of retaliation decision. The decision to use U.S. nuclear forces would be transmitted through communications links that rely on both commercial and military cable, radio, and satellite systems.

Decision evaluation, force retargeting, escalation control, and conflict termination. Once the order to retaliate has been given, feedback to the NCA is necessary to determine the effects of the decisions taken (for example, the damage inflicted on the attacker). From this information forces might be retargeted, the degree of nuclear escalation would be determined, and communication with allies and adversaries would take place to end the conflict. To carry out these tasks the C^{3}I system would have to be able to function properly in a nuclear environment, not merely survive an initial attack.

The C^{3}I system also needs to be credible in the eyes of a potential

32. Note that by a 1947 act of Congress the line of succession if the president and then the vice president die or become disabled is the Speaker of the House, the president pro tempore of the Senate, the secretary of state, the secretary of the treasury, the secretary of defense, the attorney general, the postmaster general, and then the secretaries of agriculture, interior, commerce, labor, and so on in the order of the date of establishment of their departments. The secretary of defense serves as deputy commander in chief of the armed forces in peacetime. The precise order of succession within the NCA concerning the responsibility to authorize nuclear weapon use in the event both the president and vice president are incapacitated is, however, not public knowledge.

attacker; credibility of the system would, of course, imply that the United States could retaliate as its leaders saw fit. Moreover, ideally the system should be flexible, reliable, and responsive even under the most adverse nuclear conditions. It has to be assumed that the Soviets would try to prevent the American C^3I system from functioning in a variety of ways: physical destruction of its components; use of electromagnetic pulses and other means to interfere with communications; adoption of electronic countermeasures (C^3CM) to confuse the system; and sabotage. It has been argued that the Soviet Union would not want to destroy the American C^3I system for fear of a spasm, or all-out, American response. But available information about Soviet doctrine and targeting capabilities suggests that the Soviets do not share this view, and American planning with respect to C^3I is quite correctly concerned with the system's vulnerabilities.

A great deal of money has been allocated recently to increasing the number and survivability of sensors, reducing the vulnerability of existing command centers to countermeasures, enhancing communications between the command posts and the nuclear forces (including the addition of an extremely low-frequency [ELF] system with radio antennas in Michigan and Wisconsin to improve communications with the SSBN fleet), and accelerating development of an enduring C^3I system.

The effectiveness of the current C^3I system in a protracted nuclear war would be doubtful. A study group of which I was a part concluded: "Survival for weeks or even months might be possible in exchanges limited to tens of weapons; in exchanges of hundreds of weapons, survival might be limited to hours or perhaps days. It is unreasonable to expect the C^3I system to survive beyond an initial response when confronted with an attack involving thousands of weapons."[33]

Since most scenarios involving important counterforce targets (for example, ICBM silos, SSBNs in port, and strategic bombers) call for several thousand weapons, it is unlikely that forces could be kept in strict control under such circumstances. Strengthening the present C^3I system should therefore be among the highest priorities for American strategic planners. With contemporary technologies, however, it is doubtful that the system could be strengthened enough to endure in the face of a sophisticated nuclear attack.

33. See Barry M. Blechman, ed., *Rethinking the U.S. Strategic Posture: A Report from the Aspen Consortium on Arms Control and Security Issues* (Ballinger, 1982), pp. 267–68.

Civil Defense

Finally, there is the question of how feasible it is to protect civilians. Civil defense has not been supported either financially or doctrinally since the early 1960s. But in the early 1980s it once again emerged as a major aspect of the American strategic debate, after revelations that first appeared in 1976 and 1977 about the extensive Soviet civil defense program.[34] Since American attitudes about civil defense are strongly influenced by judgments about the magnitude and capabilities of the Soviet civil defense program, one should review what is publicly known about it.[35] And because of the controversy surrounding its potential significance, the Soviet program needs to be evaluated on several grounds: its place within the Soviet defense establishment; its physical parameters; its potential effectiveness; its role in Soviet strategy; and the implications of the program for American defense policy.

The Soviets have tried almost continuously since the mid-1950s to protect citizens and heavy industry from potential nuclear attack. Because of the awesome destructive power of nuclear weapons, however, and particularly because of the absence of ballistic missile defenses, most of the American defense community has believed that the Soviet Union would not be able to prevent the United States from delivering a retaliatory blow so devastating that the Soviet Union would cease to exist as a modern society. But as the magnitude of the Soviet program has become better understood, this assumption has been challenged.

Available evidence suggests that civil defense is a high priority of the Soviet leadership. Responsibility for civil defense was transferred from the Ministry for Internal Affairs to the Ministry of Defense in 1961 and was placed under the leadership of a well-known military figure, Marshal V. I. Chuikov. Shortly after Marshal Chuikov took command, civil defense was ranked organizationally at the same level as the strategic rocket forces, the army, navy, and air force. In 1972 the program was

34. See particularly the congressional testimony by Leon Goure and T. K. Jones in *Civil Defense Review*, Hearings by the Civil Defense Panel of the Subcommittee on Investigations of the House Committee on Armed Services, 94 Cong. 2 sess. (GPO, 1976), pp. 187–282. An additional useful reference is *Civil Preparedness Review*, pt. 2: *Industrial Defense and Nuclear Attack*, Joint Committee Print, Report by the Joint Committee on Defense Production, 95 Cong. 1 sess. (GPO, 1977), pp. 55–100.

35. A dated but useful source remains *Soviet Civil Defense* (Washington, D.C.: Director of Central Intelligence, July 1978).

turned over to Colonel General Aleksandr T. Altunin, who at that time was also named a deputy defense minister. Altunin, born in 1922, is judged to be a military figure of considerable stature, charisma, and promise. In February 1977 he was promoted to the rank of full general. Moreover, according to recent estimates, Altunin oversees more than 100,000 full-time government personnel, including a staff of more than 50 senior generals posted throughout the country. From an organizational standpoint, therefore, the Soviet civil defense program is well positioned and highly valued.

The physical parameters of the Soviet civil defense program cannot be stated with certainty, for three reasons. First, from 1971 through 1975 U.S. intelligence did not focus on Soviet civil defense efforts, partly because most American analysts did not think they were strategically important and partly because U.S. intelligence had limited resources. Information on Soviet civil defense was not sought through interpreting satellite photographs, monitoring communications, or interrogating people. Only in 1977 did the intelligence community begin to review existing data and get more data to determine the dimensions of the program. Second, many assertions about Soviet civil defense are contained in Soviet civil defense manuals. It is difficult to know whether these statements reveal what the Soviets would like to achieve or what they have already achieved. Third, important aspects of what the United States knows about the Soviet program are classified to protect sources and intelligence methods.

It is generally accepted that the Soviet program is a steady and extensive set of activities designed to protect as much as possible political and military leaders, industrial facilities, and food supplies.

Measures to protect Soviet leaders seem to be far more elaborate than those the United States has adopted to protect its leaders. The Soviets have allegedly built several hardened, dispersed, and redundant facilities including important command, control, and communication centers designed to withstand nuclear attacks. There are hardened headquarters throughout the Soviet Union, Eastern Europe, and Mongolia, which, most notably, include facilities within eighty miles of Moscow for the first echelon of the Soviet government and armed forces. The Soviets have hardened many other installations, including missile silos, nuclear storage bunkers, hangarettes, and bunkered surface-to-air missile launchers.

The Soviets apparently plan to protect civilians by dispersing them and housing them in shelters. Shelters have been built for major apartment complexes in Moscow, Leningrad, and Kiev, and more shelters are being built in basements of buildings or in separate facilities. Existing facilities that could be used as shelters, such as the Moscow underground, are also important in Soviet civil defense.

Plans call for the evacuation of Soviet cities within seventy-two hours. Evacuees are to go to collective farms, where, presumably just before a war begins, they will build primitive shelters for themselves and, if possible, begin to farm. This plan is intended to provide the best possible protection against nuclear, chemical, and biological weapons. To facilitate the functioning of this system, a network of local, regional, and national civil defense organizations has reportedly been created. It is estimated that this network provides each schoolchild with more than 100 hours of instruction in the effects of nuclear weapons and in civil defense measures. U.S. intelligence does not know whether Soviet leaders, in the event of war, would order the population to evacuate the major cities or have people use existing shelters, or both.

Some analysts have argued that the effectiveness of a system based on population dispersal is doubtful, since the system has not been widely tested. A rebuttal to this argument, however, is the experience in Phnom Penh, Cambodia, shortly after the communist takeover in 1975, when the entire city was evacuated without previous tests on a few hours' notice. Although the dispersal of people from Soviet cities would be a provocative act certain to stimulate an immediate increase in the readiness of U.S. forces worldwide, it is not at all clear what other American responses—diplomatic communications, movement of civilian populations, or preemptive use of military force—would be appropriate.

The Soviet Union has also established industries in new towns in remote areas and taken steps to protect both the facilities and the workers from the effects of nuclear weapons. The establishment of new industrial towns as a civil defense measure can be viewed with considerable skepticism, however. First, these towns make good economic sense anyway, because they bring factories closer to the site of natural resources and thereby greatly reduce transportation costs. Second, the older industrial facilities have not been dispersed and remain prime targets in the event of war. Third, many of the new facilities, such as the Kama River Truck Plant, are quite huge and are easily targetable by an adversary bent on destroying Soviet industry.

Nonetheless, the Soviets have not only built industrial facilities in remote locations but have also reportedly built blast shelters in some of the major manufacturing plants and have developed techniques to protect heavy machinery from shock and other weapon effects. These techniques include placing machines on mounting slabs, spraying them with protective grease, surrounding them with sandbags, and filling cavities with water to equalize the blast overpressure so that it would act equally on all sides of each machine component.[36] There is no evidence, however, that the Soviets have sought to protect vital industrial facilities such as petroleum refineries or electric power plants.

Finally, the Soviets have allegedly built underground storage bunkers for grain and possibly other foodstuffs. The extent of this stockpiling is uncertain, and it may be that these reserves would be used for the Soviet military rather than civilians. Apparently, these reserves are untouchable in peacetime. Consequently, when crop failures have led to grain shortages in the past, the Soviet government, rather than tap the reserves, has purchased grain from the United States, Canada, and Australia and has ordered the slaughtering of livestock.

How effective these measures might be is the subject of considerable debate. Some analysts argue that the Soviet civil defense effort has led to the development of a "war survivability gap"—that the Soviet Union would survive a nuclear war far better than the United States would. It has been claimed that in a nuclear war between the two powers, ten Americans might perish for every Soviet death (indeed, there are some who claim that the ratio would be perhaps as great as forty to one) and that the ability of the Soviet Union to recover and resume functioning as a modern, industrialized society is enhanced by its civil defense system.

There are reasons to be skeptical of this judgment, however. First, fatality ratios will depend on the nature of the conflict, the speed of the evacuation of Soviet citizens from the cities, and countermeasures implemented by the United States. Second, hardened facilities that could be identified would be targeted. Assuming that the United States had enough retaliatory forces to strike them, these facilities would fail to survive a war. Third, in the event of a Soviet first strike, the United

36. The Boeing Company conducted a feasibility study using these techniques and found that machinery protected in such a manner could withstand blast overpressures of up to 300 psi compared with about 10 psi for unprotected machinery. The Boeing Report is an appendix to *Defense Industrial Base: Industrial Preparedness and Nuclear War Survival*, Hearings before the Joint Committee on Defense Production, 94 Cong. 2 sess. (GPO, 1977), pt. 1, pp. 55–133.

States would surely use its retaliatory forces at least partly to destroy Soviet transportation networks, which are far more primitive than those in the United States, and known food supplies. Mass starvation might well be the fate of many Soviet citizens who survive a nuclear exchange. Fourth, the United States could deploy and use highly radioactive weapons to inflict very high levels of radiation on the Soviet homeland. Fifth, the United States could alter its nuclear war plans so that sea-based warheads would be delivered at intervals timed to keep the Soviet population evacuated and Soviet industrial production near zero. By programming U.S. forces to carry out repetitive attacks over many months, regardless of the degree of damage done to the United States, the United States could prevent the Soviet Union from recovering.

Soviet strategic doctrine indicates that war survivability is a central element of Soviet strategy. However one assesses the significance of Soviet strategic writings, it seems clear that Soviet leaders endorse the concept of damage limitation, that is, the ability to minimize the damage to the nation should war—even nuclear war—occur. This is a strategy that the United States decided in the early 1960s was provocative strategically and not feasible economically, politically, and technologically.

To be sure, there are other plausible explanations besides strategic doctrine for Soviet behavior: a preoccupation with defense because the people have suffered from war so severely in the past; a rigid bureaucratic structure that, once authorized to promote a civil defense, has developed a program far more extensive than Soviet leaders originally intended; a preoccupation with the Chinese threat, to which civil defense efforts might be directed; and a way to control citizens in wartime.[37] In any case, the Soviet civil defense program, coupled with the advanced offensive weapon systems now being deployed by the Soviet Union, is a potential threat to strategic stability.

Revelations about Soviet civil defense strengthened the arguments of Americans who wanted to deploy weapon systems that could destroy every hard target in the Soviet Union. These revelations also created renewed support for a U.S. civil defense program. Indeed, the American

37. Students of Soviet and Russian military history have noted that the government has had a chronic problem in retaining the allegiance of significant numbers of its citizens under war conditions and that placing them in well-established civil defense facilities could be a way to ensure that the society would still function according to the leadership's requirements.

response illustrated how the enhanced defense of one superpower could produce anxiety among the leadership and citizens of another. (This is a point expressly denied by President Reagan in his support for SDI.) For Soviet civil defense has weakened, at least marginally, the confidence of some analysts in the capability of the U.S. deterrent force and intensified their concerns about Soviet intentions.

The American civil defense bureaucracy was reorganized by the Carter administration through the establishment of the Federal Emergency Management Agency (FEMA). The Reagan administration went beyond Carter by seeking funds for a program that would include:

—crisis relocation of 150 million Americans from 400 potential target areas to the countryside;

—fallout protection, involving the construction of millions of shelters to protect the U.S. population from radioactive fallout;

—blast shelters that would protect 4 million "essential" workers at their workplaces from the immediate effects of nuclear explosions;

—industrial protection through the dismantling, dispersal, and burial of machinery before a nuclear attack; and

—continuity of government through the provision of hardened facilities and duplicate records for thousands of key government officials so that they could continue to perform their duties and, through specific plans, protect presidential successors.

It is uncertain whether this program would accomplish its goals, strengthen strategic stability, and be worth its cost (at least $10 billion over seven years), but it would probably not achieve any of these.[38] It is unlikely that such a program would accomplish its goals because a deliberate Soviet attack on the U.S. homeland would presumably take into account dispersal patterns of an evacuation program. Since only a fraction of the Soviet ICBM force is needed to strike at U.S. retaliatory forces, the Soviet Union would have large numbers of ICBMs and other warheads available for striking hardened U.S. evacuation posts, food and water supplies, medical facilities, transportation networks, industrial sites, and the like.[39] Strategic stability would be marginally reduced.

38. The most persuasive case in support of a program of urban evacuation and population dispersal has been made by Samuel P. Huntington. See his statement in *Civil Defense*, Hearing before the Senate Committee on Banking, Housing, and Urban Affairs, 95 Cong. 2 sess. (GPO, 1979), pp. 24–45.

39. In an attack involving several thousand nuclear weapons detonated on the American homeland, the long-term biological consequences are extraordinarily dim. There would be extended periods of below-freezing temperatures, low light levels, high

The implementation of a complex civil defense network would require Soviet planners to revise their targeting options and would probably stimulate their offensive force deployments. It seems reasonable to conclude that the money should be spent on improving strategic forces to deter attack rather than on improving defense to ostensibly limit damage.

Three conclusions drawn from this chapter have, ironically, salutary effects: first, we have not suddenly shifted to a war-fighting doctrine—these plans have been in effect for many years; second, the inability to make a plausible case for civil defense is a stabilizing development; and third, that the C^3I systems of both countries might not function well in a nuclear environment should give pause to those advocating the feasibility of conducting protracted limited nuclear war.

Nonetheless, high-accuracy weapon systems, confusing and provocative strategic doctrines, and vulnerable command and control facilities do not bode well for U.S. survival after the start of a nuclear war. So it is understandable that Americans should turn to arms control to try to alleviate their difficulties.

---

doses of ionizing and ultraviolet radiation, probable disruption of photosynthesis by the attenuation of incident sunlight, and contaminated or destroyed agricultural and aquatic ecosystems. For those who initially survived the attack the prospect of enduring in a cold and dark environment for a protracted period would also have terrible psychological effects. See Carl Sagan, "Nuclear War and Climatic Catastrophe," *Foreign Affairs*, vol. 62 (Winter 1983–84), pp. 257–92. These findings, commonly known as "nuclear winter," suggest that an attacker could be destroyed by the effects of his own attack, even if there was no retaliation in kind. While the technical basis of this conclusion is the subject of considerable debate, it reinforces the general perception that nuclear weapon use is fraught with danger and disaster—a useful reminder to political leaders.

CHAPTER SIX

Arms Control as a Means to Threat Control

If a man will begin with certainties he will end with doubts, but if he will be content to begin with doubts he shall end in certainties.

Francis Bacon, *Advancement of Learning*

ARMS CONTROL has fallen on hard times. After more than a dozen agreements reached between 1959 and 1974 by the United States and the Soviet Union—many involving scores of other countries—no significant accords have been legally implemented in the last decade. In spite of the agreement reached at Geneva by the United States and the Soviet Union in January 1985 to resume nuclear arms control negotiations in a timely fashion, the whole arms control process continues to be surrounded by skepticism, pessimism, and disillusionment. One wonders how much longer the superpowers will have the political will to send their representatives to the negotiating table only to have them return home empty-handed.

The reasons for this decline, I believe, are that the objectives of arms control have been ill-defined and that consequently inappropriate criteria have been used to evaluate its effectiveness. There are three different perspectives on the purposes of arms control. One sees arms control as it was defined more than twenty years ago, as consisting of "all the forms of military cooperation between potential enemies in the interest of reducing the likelihood of war, its scope and violence if it occurs, and the political and economic costs of being prepared for it."[1] These might be called the classical objectives of arms control, which, as interpreted by several administrations, have essentially been approached through negotiated arms control agreements between the two superpowers. Of this approach former Secretary of State Cyrus Vance wrote:

1. Thomas C. Schelling and Morton H. Halperin, *Strategy and Arms Control* (New York: Twentieth Century Fund, 1961), p. 2.

Our security must be genuinely advanced by vigorous pursuit of arms control agreements—comprehensive, far reaching, balanced, equitable, and verifiable. . . . Neither [superpower] can seek a decisive nuclear advantage without the risk of provoking an attack in which both would be destroyed. . . . their security must be based on an unparalleled degree of cooperation. It must be common security.[2]

A second perspective has involved using arms control as an instrument of American foreign policy, with the explicit purpose of enmeshing the Soviet Union in a web of international relationships that would moderate the aggressiveness of its intentional behavior. Linkage between military restraint and economic assistance was the centerpiece of détente. A third perspective countered that the American democratic system cannot engage in arms control without being lulled into a false sense of security. Supporters of this view have believed that arms control is harmful to American national interests and either should be linked more explicitly to defense planning or abandoned altogether.

What the actual record demonstrates, however, is that arms control is really threat control, that it concerns the quest for unilateral gains by each party as much as, say, labor-management negotiations, and that it will succeed only if each side sees that the process of arms control reduces security threats in ways that no other method or strategy can.

The Classical Objectives

The traditional notion of arms control stemmed from the writings of nongovernment defense intellectuals in the late 1950s and early 1960s. These analysts were looking for a constructive middle ground between utopian and what they felt to be dangerous notions of unilateral nuclear disarmament on the one hand and, on the other hand, the accumulation of thousands of nuclear weapons and delivery vehicles in an open-ended nuclear arms race with the Soviet Union.

At the time it was recognized that Soviet and American nuclear forces were designed so that the side striking first had a substantial advantage over the side under attack. Arms control advocates called for measures that would minimize the difference in effect between first-strike and second-strike capability, hoping to strengthen the deterrent of the initial attack and thereby enhance the stability of the strategic balance.

2. Cyrus Vance, *Hard Choices: Critical Years in America's Foreign Policy* (Simon and Schuster, 1983), pp. 417–18.

The Kennedy administration and its successors adopted this reasoning, which is a fundamental principle of SALT, at least from the American perspective.[3] Indeed, the enthusiasm to begin the SALT negotiations that was generated in the Johnson administration in 1967–68 can be traced directly to the initiatives of Secretary of Defense McNamara, who hoped to use the negotiating process as a way to curtail the deployment of Soviet and American antiballistic missile systems. McNamara interpreted the deployment of these systems as "destabilizing" because, among other things, they would give a prospective attacker confidence that a retaliatory attack could be absorbed successfully, thus increasing the attacker's incentive to strike first.

As the Soviet Union achieved strategic parity with the United States, the need to maintain parity became a principal U.S. objective in the SALT negotiations. Paul Nitze, the Defense Department's senior representative in the U.S. SALT delegation from 1969 through mid-1974, stated that the United States had three objectives: to seek both the reality and appearance of essential equivalence in the permitted levels of strategic arms of both sides; to seek limitations that would help maintain crisis stability and thus reduce the risk of nuclear war; and to find a way to deescalate the arms competition. Nitze had made it clear that the third objective had lowest priority, the second was of the greatest strategic value, and the first was vital because of domestic political considerations that dictated that the United States not be perceived as slipping behind the Soviet Union in strategic power.[4] At least through the Nixon and Ford administrations, neither economic savings nor disarmament were given high priority.

It is useful to examine how successfully the SALT I agreements met these objectives. The ABM Treaty and its subsequent modifications helped reduce the potential difference between the damage that could be inflicted by a first strike and a retaliatory attack. In this sense the treaty supported the objective of strategic stability. Moreover, the budgetary savings resulting from halting the deployment of ABM systems must also be credited to the treaty—a bonus in view of the low priority given this objective. However, the treaty runs directly counter to the objective of reducing damage should war occur. In addition, it is unlikely

3. Arms control has of course not been limited to bilateral U.S.-Soviet nuclear arms talks, but these have been the centerpiece of arms control activity for fifteen years.

4. Paul H. Nitze, "The Strategic Balance between Hope and Skepticism," *Foreign Policy*, no. 17 (Winter 1974–75), pp. 138–39.

that the threat of nuclear war would be much greater today had no treaty entered into force, given the uncertain technical feasibility of ABM systems for either city or silo defense and the fact that such systems could be quite easily saturated by an attacking force at low cost.

The SALT I Interim Agreement partially satisfied Nitze's third objective of placing limits on certain aspects of the arms competition, namely the number of ICBM and SLBM launchers. And because in May 1972 the Soviet Union was still deploying these launchers while the United States had no new launcher deployment program even planned, the case can also be made that the Interim Agreement froze a situation that would otherwise have continued to shift in favor of the Soviet Union. Nonetheless, the numerical superiority the agreement granted to the Soviets was clearly troublesome to Americans and was accepted by the U.S. Senate largely because it was thought to be a transitional step toward a more equitable and permanent limit on offensive forces. In retrospect, the inability of the Interim Agreement to limit either the modernization of existing weapon systems—particularly the number, yield, and accuracy of MIRVs—or the introduction of new systems leads to the conclusion that the agreement satisfied none of the objectives of U.S. SALT policy.

The achievement of numerical parity and then its rapid disappearance because of Soviet force acquisitions became a topic of political debate, even though numerical comparisons are not necessarily valid indications of military advantage. The numbers nonetheless became politically significant. Since the Kremlin was unwilling in the 1970s to reach SALT agreements that would constrain its nuclear weapon modernization program (it had, after all, no incentive to do so), the strategic nuclear balance shifted in numerical terms in the 1970s, and it was clearly demonstrated that such shifts have potent political and psychological effects.

Linkage as an Objective

The Nixon-Kissinger approach to arms control, which continued under President Ford, rested on attempts to constrain the Soviet Union's efforts to exploit its burgeoning military power. It is doubtful whether Nixon and Kissinger ever seriously believed that the SALT process could prevent the Soviet Union from becoming a military superpower, given the Soviet Union's intention to achieve this status. Rather, SALT was seen as one tool for gaining leverage over some aspects of Soviet foreign policy. The objective, apparently, was to enmesh the Soviets in a complex network of international relations so that over time Soviet leaders would adopt policies that were less hostile to and more cooperative with the United States in particular and Western nations in general.[5]

Because Nixon and Kissinger evidently believed, at least in the short term, that the fine details of the Soviet-American nuclear balance were neither militarily nor politically important, they were willing to forge arms control agreements with the Soviet Union that would grant it selected numerical advantages, as long as these advantages could be offset by superior American technology—particularly highly accurate MIRVs. Such agreements were to be part of a package that would eventually include significant American aid to the Soviet Union: improved terms of trade and credit; wheat and other foodstuffs; high-technology equipment and devices, particularly digital computers; training in the application of advanced management techniques; and aid in the development of Soviet natural resources. Successful arms control agreements were viewed as a necessary precondition for this kind of assistance to the Soviet Union on a sustained basis.

Using this approach, Nixon and Kissinger hoped to promote "linkage," whereby Soviet decisionmakers would not want to advance policies antithetical to American foreign policy for fear of risking the economic aid on which they had come to depend. Help in extricating the United States "properly" from Vietnam and legitimate support for peaceful solutions to the conflict in the Middle East exemplified the Soviet responses that this approach was intended to induce. Although it was certainly true that Nixon and Kissinger hoped to limit the growth of Soviet military programs through SALT, it was the broader objectives of complicating Soviet decisionmaking and inhibiting aggressive Soviet foreign policy that were at the heart of their strategy.

There are three points about the strengths and weaknesses of the linkage concept and détente that are important to mention. First, the Soviet Union resisted manipulation more than the designers of the policy probably expected it would. Soviet leaders refused to be enticed into a foreign policy more in line with American aims and desires. The Soviets

5. A brief but useful summary of this approach may be found in Helmut Sonnenfeldt, "The Meaning of Détente," *Naval War College Review*, vol. 28 (Summer 1975), pp. 3–8.

distrusted the political strings attached to American economic aid, especially in connection with the Jackson-Vanik Amendment that linked Soviet most-favored-nation status to Jewish emigration. They chose to reject the aid rather than accept the strings. The designers of détente proved to be more its captive than was the Soviet leadership, and the anticipated complication of Soviet decisionmaking failed to materialize in any meaningful way.

Second, the policy suffered from having been conceived and executed principally by Nixon and Kissinger. Their tendency to dominate the policy created enormous resentment among members of the bureaucracy, who were neither included in the formulation of the policy nor given any credit when it seemed to be working successfully in 1972–73. Ultimately only Nixon, Kissinger, and eventually Ford had a stake in détente within the government. By failing to build bureaucratic bridges, they left themselves open to attack from within when it became evident that success was not inevitable. In the defense sector, the inability of either the SALT process in particular or the détente policy in general to slow the pace of Soviet military deployments pushed the secretary of defense and the Joint Chiefs of Staff from an initial position of healthy skepticism to a stance publicly critical of détente.[6]

Third, in Nixon and Kissinger's effort to make SALT conform to overriding political and foreign policy objectives, it became increasingly divorced from both defense policy and the well-established goals of arms control. The drafting of ambiguously worded agreements, the reliance on unilateral statements that could be ignored by the Soviets and exploited by domestic opponents of SALT, the emphasis on secrecy, and the overselling of the agreements beyond their intrinsic worth contributed to a growing skepticism about the value of the SALT negotiations. Although Nixon and Kissinger thought the process of negotiation was extremely useful, it became increasingly difficult to sustain it once the results were judged to be of dubious value.

After the SALT I agreements were ratified, Soviet nuclear force deployments and Soviet policy in developing countries undermined

6. As support for détente began to deteriorate during the Ford administration, Secretary of Defense Rumsfeld observed: "Détente needs to be understood for what it is: a word for the approach we use in relations with nations who are not our friends; who do not share our principles; who we are not sure we can trust; and who have great military power and have shown an inclination to draw on it." See *Annual Defense Department Report, Fiscal Year 1977*, p. 9.

assumptions about the cooperative potential in the Soviet-American relationship. Soviet military and political transgressions cited in chapter 2 progressively undermined American domestic support for détente. Skepticism materialized concerning the feasibility of fashioning a relationship with the Soviet Union other than one of pure competition.

Return to the Classical Objectives

President Carter assumed office as the U.S. military posture was deteriorating and as the Soviet leadership was becoming bolder. Ironically, his initial approach to foreign policy showed ignorance of these developments and was designed mostly to be in contrast with that of his immediate predecessor (an example of the "clean slate" phenomenon cited in chapter 3). Foreign policy under Carter initially emphasized relations with Western Europe and Japan and less preoccupation with the great communist powers; global issues such as nuclear proliferation and conventional arms transfers; commitment to human rights as a foreign policy goal; the nonmilitary aspects of international security, particularly international economic affairs, rather than balance-of-power diplomacy; and openness and multiple participation in foreign and defense policymaking as opposed to secrecy and solo performances.

The Carter administration's approach to SALT, seen in this context, was a return to the classical objectives of arms control. Linkage as a way to cope with Soviet military power was jettisoned. Carter apparently sought to test whether the Soviet Union was fundamentally interested in arms control, and he was willing for quite some time to forgo the completion of SALT agreements unless they helped stabilize the strategic balance. This is how he expressed his position:

> Our view is that a SALT agreement which just reflects the lowest common denominator that can be agreed upon will only create an illusion of progress and, eventually, a backlash against the entire arms control process. Our view is that genuine progress in SALT will not merely stabilize competition in weapons but can also provide a basis for improvement in political relations.[7]

In late March 1977 a U.S. delegation led by Secretary of State Cyrus R. Vance offered the Soviets two alternatives: completion of a SALT II agreement based on guidelines that had been worked out at the U.S.-

7. See the text of President Carter's address on U.S.-Soviet relations to the Southern Legislative Conference, excerpted in *New York Times*, July 22, 1977.

Soviet summit meeting held in Vladivostok in November 1974 but excluding limitations on cruise missiles and the Backfire bomber; and the preferred American approach, a "comprehensive" proposal that departed significantly from the Vladivostok guidelines. The stated objectives of the comprehensive proposal were to provide both sides with political parity and strategic stability, with emphasis on constraining those aspects of each side's strategic programs that were seen as most threatening to the other. The Soviet Union rejected both proposals, claiming that they departed from the Vladivostok guidelines, that they were one-sided in favor of the United States, and that they were especially unsatisfactory with regard to the freedom to deploy long-range cruise missiles.

Two years later, however, President Carter and Leonid Brezhnev signed the SALT II Treaty in Vienna, which largely conformed to the Vladivostok guidelines with respect to ICBM and SLBM limitations and permitted a Soviet advantage in heavy ICBMs. The treaty was severely criticized in the United States because of the constraints it imposed on cruise missile deployments in Europe and its lack of clarity about Backfire deployment limitations, as well as the questionable verifiability of some of its provisions.[8]

The Soviet Union's objectives are obviously also highly relevant to an evaluation of the SALT process. Official Soviet statements have consistently emphasized the desire to reach a condition of equality in armaments and security, to curb the strategic arms race, to adopt measures that would help avoid nuclear war, and to promote political détente between the superpowers. But these broad principles do not reveal the fundamental political and military aims of Soviet SALT policy. These aims can only be inferred from Soviet writings, military deployments, political behavior, and the observations of a few former Soviet citizens and officials now living in the West. Such inference is unavoidably distorted by the perceptions of the analyst, which are, in turn, strongly influenced by personal values and judgments. The following comments on Soviet SALT objectives are offered with these difficulties in mind.

Soviet SALT policy was guided from its inception by a small group of officials led by General Secretary Leonid Brezhnev. During Brezhnev's term in office beginning in 1964, the Soviet Union became a

8. The controversy surrounding the terms of the agreement is admirably reviewed in Strobe Talbott, *Endgame: The Inside Story of SALT II* (Harper and Row, 1979).

superpower and acquired more strategic weapons than the United States had. During this period Brezhnev and his colleagues had to grapple continuously and conspicuously with the Sino-Soviet conflict and with significant economic problems. Soviet SALT policy seems to have shaped and been shaped by these two developments.

At the symbolic level the SALT I agreements codified Soviet superpower status. The agreements amounted to American recognition that the Soviet Union was a great military power, a power to be reckoned with and consulted, and to be involved in the resolution of all important regional conflicts. Whether such recognition was an objective of the Soviets' SALT policy is less important than that it became reality.

On the military level SALT failed to constrain Soviet weapons programs except for ABM systems. This suited the Soviet Union and was probably a Soviet objective from the outset. A few former Soviet citizens now living in the West claim that Soviet arms control policy was never permitted to interfere with planned or ongoing military programs. The single purpose of reaching agreements, according to these observers, was to prevent the United States from deploying systems that would have been deployed in the absence of such agreements. Presumably, sophisticated Soviet analysts anticipated that the relaxation of tensions produced by the SALT agreements would strengthen the position of those Americans calling for unilateral restraint in weapon deployments. If sustained for a lengthy period, Soviet growth and American restraint would permit the Soviet Union to achieve quantitative strategic superiority while narrowing the U.S. technological lead.

The ABM Treaty did halt deployment of a marginally effective U.S. system and an ineffective Soviet system, leaving the Soviets the option of revising or abrogating the treaty if and when it could perfect a system truly capable of ballistic missile defense. The Interim Agreement met the Soviet objective of codifying numerical superiority in ICBM and SLBM launchers without interfering with Soviet modernization programs. Also, the Soviets have been able to catch up technologically with the United States in some respects—for example, ICBM accuracy—but by no means in all qualitative aspects of the strategic forces.

By remaining in SALT negotiations and maintaining relations with the United States, the Soviet Union also forestalled a rapprochement between the United States and China. A major setback for Soviet foreign policy would have been improved Sino-American relations that could have led to the marriage of Chinese manpower with American technology and sophisticated weaponry. By sustaining the SALT process and other elements of détente, the Soviets no doubt hoped to generate the complications in American decisionmaking that Nixon and Kissinger had tried to introduce into Soviet policy formulation. The Soviets might have reasoned that the Americans would hesitate to move too close to China if such a move would jeopardize strategic arms control with the Soviet Union.

Finally, the Soviet leadership hoped to benefit economically from SALT. If the United States and Western Europe could be persuaded to provide substantial economic assistance and high-technology equipment in return for a less aggressive Soviet foreign policy, the leadership would surely be willing to create the illusion of softening as long as the United States could be kept from influencing Soviet affairs. Economic assistance such as the provision of digital computer technology could also be used by the military as well as the industrial sector, thus providing an added incentive for the Soviets to behave in a way that would make them eligible for such assistance.

By signing SALT II, the Soviets at least marginally satisfied these last two objectives. Sino-American relations did not improve as much as the Soviets feared they would, more because of outstanding differences between the two nations than because of American fear of jeopardizing SALT. The Soviet Union did receive economic assistance from the West, especially through the transfer of certain technologies, although probably nowhere near the levels that optimists in Moscow must have hoped for. Also, the nonmilitary sectors of the Soviet economy seem to have received little benefit from economic relations with the West.

Overall, therefore, the Soviets had mixed success in meeting their SALT objectives. But since 1979 Soviet-American relations have deteriorated, which has stimulated American defense spending and triggered a U.S. policy of economic sanctions (except with respect to grain sales) that is jeopardizing many of the aims of Soviet arms control policy.

Criticism of SALT is justified on the grounds that it failed to meet the classical objectives of arms control or the pragmatic interests of American policymakers. The likelihood of war—even a Soviet-American nuclear war—has not seemingly decreased; the potential damage a war could inflict has not been reduced; and the resources allocated to preparation for war have not declined.[9] (Admittedly, this judgment is based on a comparison between expectations and results; it is not a

9. The ABM Treaty, it can be argued, has reduced the likelihood of war and marginally limited the resources devoted to the preparation for war while increasing the probability of damage should war occur.

comparison between how the world would have been in the absence of SALT and what we have achieved with it.) Nor has Nixon and Kissinger's objective of seducing the Soviet Union into a less aggressive stance through arms control, economic assistance, and other measures been satisfied.[10] In addition, the major assumptions that provided the basis of American support for arms control with the Soviet Union in the late 1960s have been severely shaken. In the 1980s most Americans believe that their relationship with the Soviet Union is fundamentally competitive. They feel that numbers of weapons do matter and that they do not wish to come in second in any significant comparison with the Soviet Union. Americans see a dangerous and turbulent world in which military force will play a key role. But they are also hopeful that negotiated arms control can limit bilateral nuclear arms competition.[11]

The Lulling Argument

Since the Soviet invasion of Afghanistan prompted President Carter to ask the Senate to delay consideration of the SALT II Treaty, the role of negotiated arms control as an element of U.S. national security policy has undergone a difficult reappraisal. The Reagan administration reformulated SALT as START (strategic arms reduction talks) and began to seek major reductions in key indicators of Soviet nuclear strategic might.

The Reagan administration's arms control approach is based on three criticisms of SALT. First, the SALT agreements had failed to halt the buildup of Soviet nuclear forces; that buildup gave the Soviet Union nuclear superiority over the United States. Second, the SALT process had lulled Americans into underestimating the Soviet military threat and the military forces and budgetary resources that the United States requires to meet this threat. Third, SALT agreements have not solved

10. It should be noted, however, that the Jackson-Vanik Amendment, linking most-favored-nation status for the Soviet Union with the fate of Jewish emigration, precluded a full test of the Nixon-Kissinger thesis.

11. Note that even proponents of a nuclear weapon freeze, many of whom might in fact prefer U.S. unilateral disarmament, had to rephrase their proposal to call for a *bilateral, verifiable* freeze in order to win broad public support in the 1982 general elections. For a survey of American attitudes toward the Soviet Union, arms control, and the freeze, see John E. Reilly, ed., *American Public Opinion and U.S. Foreign Policy, 1983* (Chicago Council on Foreign Relations, 1983), pp. 28–32. See also chapter 3, note 13, of this book.

and cannot solve particular military problems, such as the vulnerability of the U.S. land-based missile force.

The Reagan administration argued that arms control became divorced from defense planning, and that negotiations were pursued for the sake of reaching agreements regardless of their relations to force posture or foreign policy considerations. Because of the political support given to the SALT process, the administration claimed, national force posture decisions that would otherwise have been a natural outcome of defense planning were distorted, delayed, or nullified. SALT critics claimed that to remedy this situation it was necessary to downplay the importance of arms control negotiations, reduce their scope, lower expectations about what can be achieved through such agreements, and integrate arms control into defense planning.

However, evidence does not support the claim that arms control, even if it has failed to live up to its advance billing, has harmed the U.S. military position. In the 1970s several programs evolved that played critical roles in shaping the U.S. nuclear force posture: ABM systems, MIRVs, the Trident submarine, the B-1 bomber, the cruise missile, and the MX missile.

Antiballistic Missiles

U.S. defense experts have always had serious doubts about the technical effectiveness of antiballistic missiles. An ABM system could not satisfactorily defend American cities from a sophisticated missile attack because ABM systems have "leaks." Attackers could use saturation techniques, decoys, penetration aids, and other devices, or strike ABM radars first to destroy the "brains" of the system. ABM systems could possibly defend hardened targets, but to justify the systems would still be difficult, particularly in the face of their probable ineffectiveness against a coordinated Soviet SLBM pindown and ICBM attack.

However, although research and development funding for BMD systems has been cut—at least until recently—since the ABM Treaty entered into force because of congressional "arms control" sentiment, it would be simplistic to conclude that misguided notions about strategic stability, assured destruction, and arms control led Nixon and Kissinger to accept an ABM Treaty that precluded the deployment of a valuable defense system. Rather, Nixon and Kissinger wanted to halt Soviet ABM deployments, wanted a SALT agreement for domestic political purposes and as part of their strategic policy of dealing with the Soviet Union, and were willing to pay the price of sacrificing a defense system whose probable effectiveness was marginal at best.

Multiple Independently Targetable Reentry Vehicles

The decision to deploy MIRVs for both the Minuteman III ICBMs and the Poseidon SLBMs has been analyzed extensively and is the subject of considerable discussion in Nixon's and Kissinger's memoirs.[12] MIRVs have many military attractions: they increase the attacker's confidence in the ability to penetrate defended targets; they permit breadth of target coverage; and they provide the attacker with countersilo kill capability if the yield and accuracy of the reentry vehicles satisfy certain performance criteria. Also, in the 1970s they compensated the United States for having fewer launchers than the Soviet Union. MIRVs were never seriously discussed during SALT I, even though their deployment by both superpowers has been unsettling because of their ability to destroy missile silos. The fact that MIRVs *have* been deployed refutes the argument that confused American notions of strategic stability have damaged the U.S. nuclear force posture.

Trident Submarines

The Trident submarine was part of the strategic force modernization package that the Nixon administration offered in return for acceptance of the SALT I agreements. The submarine's quietness, its ability to operate deep in the ocean for long periods, and its capacity to carry larger and therefore longer-range SLBMs than its predecessor make it virtually invulnerable—the crucial characteristic of a sea-based force. However, because it carries twenty-four rather than sixteen missiles it represents in a SALT-constrained environment a reduced number of targets than would a fleet of more boats carrying fewer SLBMs. The design of the Trident submarine was dominated by power plant and crew considerations; vulnerability to enemy ASW capabilities, the principal consideration from an arms control perspective, was not a major design

12. See, in particular, Ted Greenwood, *Making the MIRV: A Study of Defense Decision Making* (Ballinger, 1975).

criterion. Hence, one can argue that the evolution of the U.S. sea-based deterrent force would not have been different without SALT.[13]

B-1 Bombers

The U.S. Air Force pushed long and hard for a penetrating bomber to replace the aging B-52 fleet, but it is stretching reality to attribute President Carter's cancellation of the B-1 program to the nefarious influence of arms control. At least two other considerations played an important role in his decision: penetrability and cost. Although the advertised warning-to-launch capability of the B-1s was better than that of the most modern B-52s, allowing B-1s to reduce the potential effectiveness of a Soviet depressed-trajectory SLBM attack against U.S. Strategic Air Command bases, the potential gains in penetrating Soviet air defenses with B-1s instead of B-52s are far less dramatic. Admittedly, such judgments are based on several technical assumptions, an ability to generate bomber corridors, and the capabilities of air defense systems. But reasonable projections about the improved effectiveness in the 1980s of Soviet air defenses and layered defenses have reduced U.S. confidence in the ability of the B-1 to hit a variety of targets deep in Soviet territory.

Also, development of the air-launched cruise missile (ALCM) confirmed the belief that a stand-off bomber could carry out many of the missions of a penetrating bomber, thus negating the need for a multibillion-dollar investment in the B-1 program.[14] There is little doubt that cost was an important element in Carter's decision. Carter ran for office on the promise of balancing the federal budget by the end of his first term. Because of the anticipated growth in the defense budget calculated early in his administration and the likely budgetary commitment to the MX, Trident, and ALCM programs, he was persuaded that funding the B-1 program would preclude a balanced budget.

Finally, the relative ease with which President Reagan resurrected the B-1 is more evidence that arms control has not dominated force posture decisions. Despite the likelihood of a new advanced technology bomber (ATB) being deployed in the early 1990s and without in any way

13. One possible caveat is that without SALT, and with considerably higher U.S. defense budgets prevailing throughout the 1970s, the navy might have pushed harder for more rapid development and deployment of the Trident I and Trident II missiles.

14. In certain contingencies the B-52 equipped with ALCMs could serve as a penetrating bomber as well.

damaging continued American compliance with the SALT I Interim Agreement and the SALT II treaty, Reagan reinstituted funding for the B-1 in his first year in office.

Cruise Missiles

The full story of the tortured development of U.S. cruise missiles is complex. What seems clear is that strong military interest in cruise missiles (ALCMs, SLCMs, GLCMs) did not materialize until after the Vladivostok summit in November 1974.[15]

It is appropriate to point out here that force posture decisions are influenced not only by available technologies, mission requirements, the politics of executive-congressional relations, congressional logrolling, and interservice rivalry, but also by intraservice politics and trade-offs. In a military organization whose reason for being and high prestige are symbolized by the ICBM, the intercontinental bomber, and the advanced fighter aircraft, there is little incentive, and indeed powerful disincentives, to wage bloody bureaucratic battles to obtain budgetary support for ALCMs—systems that would undermine the strategic rationale for the organization's bread-and-butter programs. Similarly, an organization dominated by the nuclear-powered aricraft carrier and the nuclear ballistic missile submarine has little reason to support an SLCM program that would benefit primarily the attack submarine fleet and selected surface ship programs. Finally, a fighting force organized around the soldier and armored warfare has limited interest in GLCMs, which would presumably improve the effectiveness of artillery units but would only marginally improve the service's ability to wage conventional warfare.

Since early 1975 the cruise missile in all its forms has been the subject of intense bargaining and negotiation within the United States government and in the SALT negotiations. One can conclude that U.S. interest in the cruise missile was stimulated rather than retarded by the SALT process.

15. There is some evidence to suggest that members of the Ford administration who opposed the framework of the Vladivostok agreement sought to link the cruise missile issue to constraints on the Soviet Backfire bomber. The alleged rationale was that the Soviets would not accept any constraints on the Backfire under SALT II, but desperately wished to forestall U.S. cruise missile deployment; linking the two would effectively stalemate the negotiating process, perhaps leading to the abandonment of the Vladivostok formula.

MX Missiles

The evolution of the MX land-based ICBM has been peculiar and controversial, and its outcome is still uncertain. When allegations were first made publicly in the early 1970s that the U.S. Minuteman force was vulnerable to attack, the U.S. Air Force initially mounted the most vigorous rebuttal, pointing in particular to the fratricide problem as a major obstacle for the attacker to overcome.[16] After President Carter terminated the B-1 program, the air force pushed the MX program more vigorously. It received strong support from Secretary of Defense Harold Brown and Under Secretary of Defense William Perry and the blessing of President Carter as well. It was reasoned that the MX would augment considerably the hard-target kill capability of the U.S. land-based missile force, would narrow to some extent the disparity between the Soviet and U.S. ICBM forces—perhaps thereby improving perceptions of the strategic balance—and would enable the United States to attack the entire spectrum of Soviet targets. But the ability to destroy hard targets does not fit arms control definitions of strategic stability, which urge enemies not to aim at each other's retaliatory forces. Carter decided to deploy the MX in part to gain support from Senate skeptics for ratification of the SALT II Treaty. Thus arms control considerations enhanced rather than retarded MX missile deployment.

SALT-related concerns have played a significant role in the choice of the basing mode for MX missiles, however. In 1979 members of the Defense Science Board and many other respected defense analysts endorsed a multiple vertical protective shelter system for the MX, a means of protecting the missile through deceptive basing among several shelters. Carter apparently rejected this basing mode on the grounds that it would be inconsistent with provisions of the SALT II Treaty and would not be verifiable. Instead he endorsed a horizontal basing mode that would have been more expensive and more vulnerable to attack, though more easily verifiable, than the vertical structure system. This is perhaps the first concrete example of a major nuclear force posture decision that was directly, and perhaps adversely, influenced by arms control considerations. (Both horizontal and vertical basing modes for

16. As noted in chapter 4, "fratricide" describes a condition in which the nuclear explosion of the first attacking warhead destroys the second warhead (its "brother") before the latter can strike its target.

MX were subsequently abandoned by the Reagan administration because of intense local opposition in the states where they would have been deployed.)

Conclusion

Overall, then, every administration since Nixon's can accurately claim that no important nuclear weapon program that the United States sought to deploy was constrained by the SALT I agreements, the prospect of SALT II entering into force, or the SALT process itself.

More generally, it is difficult to assess the allegation that the public and in turn Congress were lulled into a false sense of security by the rhetoric of SALT and détente and that this self-delusion undercut needed legislative support for real growth in the defense budget through much of the 1970s. Even if this happened, other considerations also affected the American defense stance. First, weakened support for defense programs resulted at least as much from the "no more Vietnams" attitude that permeated American society through most of the 1970s as from the arms control euphoria that had begun to wane by 1975. Second, the decline in U.S. economic productivity and the sharp rise in costs of imported oil since 1973 very likely influenced congressional willingness to support expensive weapons programs that were competing with several other budgetary priorities. Third, it is not at all clear that more defense funds, even if they had been forthcoming during this period, would have been allocated for strategic nuclear weapons. General purpose forces might well have been the principal beneficiary.

The perhaps uncomfortable but compelling conclusion to be drawn from this analysis is that blame for the deficiencies in U.S. nuclear forces cannot be placed primarily on arms control and its supporters. But although neither arms control nor the negotiating process has directly influenced the U.S. nuclear force posture, except in the case of the MX basing mode, arms control has by and large failed to satisfy the objectives articulated by its theorists and implementers.

Threat Control and the ABM Treaty

Because the ABM Treaty signed by the Soviet Union and the United States is such a striking exception to the general record of arms control, and because continued adherence to its terms is a major focus of public debate, it is worth considering the motivations behind this agreement in more detail.

According to the ABM Treaty, as modified by a 1974 protocol, an ABM system consists of ABM interceptor missiles, ABM launchers, and ABM radars; each side is permitted to deploy one ABM system either at its national capital or at an ICBM silo launcher site; development, testing, and deployment of sea-based, air-based, space-based, or mobile land-based ABM systems are prohibited; and "in the event ABM systems based on other physical principles . . . are created in the future, specific limitations on such systems and their components would be subject to discussion in accordance" with the articles of the treaty. The treaty is of unlimited duration and is to be reviewed by each country every five years. Each country can also seek amendments at any time through discussions at the U.S.-Soviet Standing Consultative Commission (SCC) or can withdraw from the treaty after six months' notice if it "decides that extraordinary events related to the subject matter of this Treaty have jeopardized its supreme interests."

Why did both countries agree to forgo defending themselves from nuclear attack? Each set of incentives tells us a good deal about some of the real purposes served by a negotiated arms control agreement.

The U.S. Perspective

For the United States, the ABM Treaty offered several advantages. The treaty limits Soviet ballistic missile defenses to insignificant levels, thereby assuring that U.S. strategic missiles surviving a first strike could retaliate against Soviet military, industrial, and civilian targets. In the absence of significant Soviet ABM defenses—the Soviets restricted their deployments to one obsolescent system around Moscow—U.S. strategic planners could concentrate on developing and deploying weapons that could cover a maximum number of Soviet targets. The introduction of MIRVs greatly increased the U.S. ability to strike Soviet cities and military installations.

Moreover, U.S. officials thought at the time that mutual vulnerability to retaliatory missile attack strengthened deterrence. In particular, they believed that Soviet acceptance of the ABM Treaty added validity to the concept of assured destruction, which, at the time the treaty was signed, formed the basis of U.S. declaratory policy. American officials also hoped that mutual vulnerability would reduce the incentive for either side to acquire additional offensive nuclear forces. Thus the offensive strategic arms competition would be stabilized. More important, the absence of effective defenses means that initiating a nuclear exchange or retaliating in response to an opponent's first strike would result in comparable levels of damage. Hence the incentive to strike first in a crisis would be minimal, and the crucially important goal of stabilizing the behavior of the superpowers during crises ("crisis stability") would be realized.

U.S. officials also thought that an agreement as significant as the ABM Treaty would open the way for major offensive arms control agreements calling for substantial reductions by both sides in the number of nuclear weapons and for important limitations on the characteristics of weapons still permitted to be deployed. The treaty also won the support of those concerned about the proliferation of nuclear weapons because the accord demonstrated the superpowers' commitment to arms control and disarmament, a pledge both countries had made in the 1968 Nuclear Nonproliferation Treaty.

In international political terms, this demonstration of U.S.-Soviet cooperation and of the utility of negotiated arms control agreements affirmed the détente strategy of Nixon and Kissinger. In domestic political terms, Nixon carefully orchestrated the SALT I agreements—of which the ABM Treaty is a part—to strengthen his political position. His impeccable anticommunist credentials secured him the support of conservatives in both political parties. His efforts to improve simultaneously Sino-American and Soviet-American relations were calculated to broaden his political base, secure his reelection, and codify for the historical record his credentials as a statesman of the first rank.

In budgetary terms, the ABM Treaty meant that the United States did not have to spend all the funds required to deploy an ABM system and could reduce funding for BMD research and development programs to modest levels. Indeed, some Americans applauded the ABM Treaty simply because they were skeptical about the military effectiveness of Safeguard—the ABM system developed by the United States—and thus opposed the system's deployment. A negotiated agreement in which both sides relinquished the right to deploy ABM systems would prevent the Safeguard system from being completed.[17]

17. In fact, a necessary condition for U.S. support of the ABM Treaty in 1972 was the doubt the U.S. technical community expressed about the ability of the Safeguard system or any other contemporary technology to perform effectively against a sophis-

At the same time, the treaty had certain disadvantages for the United States, some of which became more significant as U.S.-Soviet relations began to sour in the mid-1970s. Most important, the treaty simplified Soviet nuclear attack problems generally and enabled the Soviet Union to develop the capability to destroy the entire U.S. ICBM force without having to worry about penetrating U.S. defenses. Because successive administrations have failed to resolve this problem—the vulnerability of U.S. land-based missiles—the argument for modifying the treaty to permit use of BMD systems to protect the land-based missile force has gained considerable support.

In addition, the risk that the United States would become unilaterally vulnerable to attack if the Soviets developed a BMD system clandestinely and then deployed it suddenly has always disturbed some members of the U.S. defense community. The U.S. ability to verify Soviet compliance with the treaty, particularly with those terms relating to the development of new types of ABMs, has therefore been a highly charged political issue. Critics also feared that the treaty made the United States vulnerable to nuclear attacks by countries other than the Soviet Union, although this objection became less valid as Sino-American relations improved.

Those who contend that the U.S. emphasis on the doctrine of assured destruction represents an unrealistic nuclear policy have also consistently opposed the ABM Treaty. They have long argued that an ability to deter nuclear war must include a spectrum of capabilities and contingencies. These analysts insist that the ability to destroy Soviet cities in a retaliatory attack will not deter a variety of Soviet actions short of an all-out nuclear attack, including a first strike on U.S. land-based missiles. Therefore, they argue, the United States must develop limited nuclear options, including a flexible ability to target Soviet nuclear forces. Some view the treaty as a hindrance because they believe it has kept too many Americans wedded to an unrealistic nuclear weapon policy—assured destruction—that they contend the Soviets have never accepted. (As was shown in chapter 5, however, actual U.S. war plans, in contrast to declaratory policy, have long included military targets.)

---

ticated Soviet attack. During the years immediately before the signing of the treaty, experts raised several technical questions about the Safeguard system that were never adequately answered. Critics pointed to the substantial vulnerability of the system's radars and the system's inability to cope with large attacks that would saturate it. Decoys, chaff, and other penetration aids also could be used to degrade Safeguard's ability to discriminate among targets.

Some have also criticized the ABM Treaty on the grounds that as a product of the SALT process it helped lull the American public and Congress into a false sense of security. This mood resulted in insufficient modernization of U.S. nuclear forces in the 1970s. Critics maintain that inadequate funding for U.S. BMD programs has been one consequence. Technicians involved with U.S. BMD systems assert that they have been caught in a vicious cycle as a result. They need more funds to demonstrate the feasibility of developing a significant missile defense against Soviet attack, but they have not been able to obtain congressional support for more funding without first demonstrating this capability. Only with President Reagan's commitment to the strategic defense initiative has this cycle been broken.

Another concern critics voice about the treaty is that the agreement set a poor precedent for arms control because it is so technical and detailed. After the painful experience of negotiating and then failing to ratify the SALT II Treaty, many now believe that for arms control to succeed the results of negotiations must be simple and easily understood by the electorate.

The Soviet Perspective

The ABM Treaty eased the Soviets' concerns about the ability of their weapons to penetrate U.S. defenses and reach U.S. targets. Thus the treaty preserved a preemptive counterforce option (the ability to strike rapidly against the opponent's forces before they are launched) consistent with Soviet military doctrine.

The ability granted by the treaty to retaliate after a first strike has probably offered more relief to Soviet defense planners than to their American counterparts. Although the Soviet Union has made giant strides over the past few years in matching and in some cases surpassing U.S. weapons technology, most observers agree that the United States holds the edge in technologies relevant to missile defenses. Effective BMD capabilities require the most advanced technological systems that blend sophisticated sensing devices, information processors, and ballistic missiles and warheads. In combination, these systems must locate the attacking reentry vehicles, discriminate between the real RVs and their penetration aids, and finally track and destroy the RVs. Soviet defense officials may have thought in the early 1970s that the U.S. technological edge in defensive and offensive nuclear systems was so pronounced that Soviet weapons would have much more trouble penetrating a U.S. ABM system than U.S. weapons would have in penetrating a Soviet ABM system. That the United States had already deployed MIRVs on some of its missiles by 1972 while the Soviets had yet to test a MIRV probably reinforced this impression. (This Soviet concern about American technological prowess is again evident in contemporary Soviet efforts to halt through negotiated agreement U.S. military programs in space.)

The vulnerability of the United States to retaliatory attack, as codified by the ABM Treaty, must have reduced Soviet fears about the possibility of a U.S. preemptive strike in a crisis. Moreover, given the large throw-weights of the ICBMs that the Soviets were developing in 1972 and began deploying in 1975, Soviet officials may have seen the treaty as a way to channel the arms race into areas favoring them while constraining the area of greatest U.S. advantage—high-technology defensive systems.

Political considerations, however, probably dominated Soviet thinking about the ABM Treaty in particular and SALT in general. First, by linking the Soviet Union and the United States as equals, the treaties codified the superpower status of the Soviet Union, an accomplishment the Soviets had sought for its own sake and apparently for use in their relations with other nations, especially China. This achievement also probably strengthened Brezhnev's domestic political position. Second, as I have noted earlier, SALT was expected to contribute to better overall relations with the West that would be reflected in stronger economic ties and provide the Soviets with access to sophisticated Western technology. They could use such technology to benefit both their military programs and heavy industries. Third, SALT could conceivably produce over time political friction between the United States and its European allies that the Soviets could exploit to their advantage, particularly after U.S.-Soviet negotiations addressed European-based systems. Fourth, by solidifying cordial U.S.-Soviet relations, SALT could reduce the U.S. incentive to improve relations with China.

Given the momentum of Soviet offensive weapon programs during the early 1970s, it is not clear whether the Soviet leaders anticipated a major economic benefit from the ABM Treaty. But they probably expected that rubles previously budgeted for defensive strategic systems could be reallocated to offensive systems or to other military priorities. In this sense, the treaty may have provided budgetary support for the USSR's planned offensive military buildup.

Soviet leaders probably did not consider the disadvantages of the

ABM Treaty to be important. Yet they surely realized that the one system allowed them by the treaty provided only marginal defense against a sophisticated U.S. attack, especially in light of the continuing deployment of MIRVs on U.S. ICBMs and SLBMs. Probably much harder for Soviet officials to accept was that the treaty made the Soviet Union vulnerable to attacks by third parties. Unlike the United States, the Soviet Union must take seriously the military threat posed by the other nuclear powers. Soviet military planners who thought they could mount a credible defense against a Chinese, British, or French nuclear attack must have had to swallow hard when their government signed the ABM Treaty.

The treaty may have presented the Soviets with more doctrinal, political, and psychological problems than military ones. Russia, ravaged by war throughout its history, has traditionally relied on large standing armies and quantitative superiority to compensate for its persistent inferiority in weapons quality, mobility, and innovation. Emphasis on defending the motherland has deep doctrinal and psychological roots, and contemporary Soviet military and political leaders have undoubtedly inherited this feeling. To forgo a marginally effective defensive system, even temporarily, must have been wrenching for many Soviet officials.

Prime Minister Aleksei Kosygin's reaction of almost total disbelief when Secretary of Defense Robert McNamara first raised the idea of an ABM ban at the June 1967 Glassboro summit meeting was probably the typical initial view of many Soviet leaders. That Soviet leaders came to accept the merits of the ABM Treaty reflects a combination of their assessments of the advantages of the treaty, doubts about the technical merits of their own ABM system, and the staggering budgetary commitments they had already made to other areas of the Soviet military program.

The Soviets probably saw the asymmetry in the vulnerabilities of both sides' offensive missile forces that resulted from the ABM Treaty as an additional disadvantage. As noted in chapter 4, the Soviet Union has 75 to 80 percent of its total megatonnage invested in its land-based ICBM force; the United States has a much more diversified force, with only 25 to 33 percent of its megatonnage in land-based missiles. Therefore, the absence of ballistic missile defenses places a much greater percentage of strategic nuclear power potentially at risk for the Soviet Union than it does for the United States. Theoretically this asymmetry is a disadvantage for the Soviet Union, although operationally it would be of questionable significance if the Soviets perfected a substantial capability to destroy U.S. missile silos and the United States failed to develop a similar capability.

The ABM Treaty's attractiveness was bolstered by the Interim Agreement on Strategic Offensive Forces that accompanied it. This agreement granted significant, although temporary, numerical superiority to the Soviets in ICBM and SLBM launchers and in ballistic missile submarines. As a package, then, the SALT I agreements provided the Soviet Union with noteworthy military, political, and economic benefits. It must be emphasized, however, that neither superpower forswore offensive force modernization even in the absence of effective defenses.

Verification

A critical determinant of the continued feasibility of arms control is the question of verifying Soviet compliance with arms control agreements. This appears to be largely an asymmetric concern because the Soviets have raised it infrequently in the negotiating process. Presumably the openness of American society and the magnitude of the Soviet intelligence-gathering effort give the Politburo enough confidence not to be greatly concerned about Soviet ability to verify American compliance. In the American political process, however, Soviet compliance is a major issue.

If one puts aside ongoing efforts to monitor Soviet military behavior, which are carried out routinely by the U.S. intelligence community irrespective of the status of arms control agreements, the verification process involves four basic steps:

—collecting data on the status of Soviet programs;

—interpreting the data against existing treaty obligations;

—determining with a specified level of confidence Soviet compliance with existing treaties; and

—if the Soviets are judged to have violated an agreement, determining how next to proceed.

Controversy surrounds every step. Because of the ambiguous language of certain provisions of existing agreements and the inherent ambiguity in assessing Soviet behavior, honest men can differ about whether they believe treaty violations have indeed been committed. Moreover, the verification process is itself not immune to political pressure on the part of individuals and institutions. There can therefore be a good deal of disagreement over the judgments that the verification process yields.

If Soviet violations are indeed codified, several avenues then need to be explored. If the violation is not judged to be militarily significant (itself a contentious matter), no follow-up action may be taken, although the transgression would surely claim the attention of senior-level government officials. If, however, the violation is judged to be significant, several options are available: raise the issue formally in the Standing Consultative Commission (SCC), which was established at the time of SALT I expressly to hear such grievances; raise the issue through other diplomatic channels such as directly with the Soviet ambassador to the United States or, through the U.S. ambassador in Moscow, with officials in the Soviet Foreign Ministry; leak it to the press and have the issue enter the American public debate "unofficially"; or go public through formal statements by the president or other high-level U.S. officials.

At present there are many outstanding issues regarding Soviet compliance with both the SALT I and SALT II agreements (note that while SALT II never legally entered into force, both the United States and the Soviet Union have agreed to abide by its provisions since 1979). Particularly disturbing are allegations that the Soviets are constructing a large phased-array radar at Krasnoyarsk in Central Asia that would almost certainly violate the ABM Treaty.[18] According to a report submitted by the Reagan administration to Congress in January 1984:

> The Soviet Union is violating the Geneva Protocol on Chemical Weapons, the Biological Weapons Convention, the Helsinki Final Act, and two provisions of SALT II: telemetry encryption and a rule concerning ICBM modernization. In addition, we have determined that the Soviet Union has almost certainly violated the ABM Treaty, probably violated the SALT II limit on new types, probably violated the SS-16 deployment prohibition of SALT II, and is likely to have violated the nuclear testing yield limit of the Threshold Test Ban Treaty.[19]

18. See White House, Office of the Press Secretary, "The President's Report to the Congress on Soviet Noncompliance with Arms Control Agreements," January 23, 1984, pp. 3–4.

19. Ibid, p. 1. Subsequent to this publication the Reagan administration released the report of the General Advisory Committee on Arms Control and Disarmament, "A Quarter Century of Soviet Compliance Practices under Arms Control Commitments: 1958–1983" (Washington, D.C.: The Committee, October 1984). This report cites a long list of Soviet actions considered as material breaches of arms control agreements. These are divided into the following categories: violations of international obligations, breaches of authoritative unilateral commitments, and circumventions.

Although neither SALT II nor the Threshold Test Ban treaty has been ratified by the U.S. Senate, they are part of the existing arms control regime that is at the core of contemporary U.S.-Soviet relations. With the publication of more detailed evidence supporting these claims, it may become increasingly difficult to sustain American political support for further arms control negotiations. Critics of the administration will claim that this is precisely the intent behind the publication of this material. It must be recognized, however, that the accumulation of numerous incidents of questionable Soviet behavior—which have in fact taken place—puts even the most ardent supporters of arms control in a precarious situation. If the issues are raised at the SCC and resolved satisfactorily to all parties, no harm is done. But since this has not happened in several instances, Washington has had the choice of suppressing the information, in which case it would be accused of being a witting accomplice to Soviet transgressions, or revealing the violations and running the risk of doing further damage to U.S.-Soviet relations. There is no simple escape from these difficult choices. The ease of the verification task, it should be emphasized, is directly correlated with the clarity of the treaty language specifying what behavior is permissible. Vague and complex language is the enemy of effective verification.

Arms Control Options

Given this evolution of the arms control experience and the particular difficulties raised by the verification issue, what options are available? Several have been raised in the public debate, and a few additional proposals are worthy of consideration.

A Nuclear Freeze

This proposal has commanded widespread public attention and support and was endorsed by presidential candidate Walter Mondale. It has been most often proposed as a mutual and verifiable ban on the production, testing, and deployment of all new missiles and aircraft that have nuclear weapons as their principal or sole payload. Advocates of the freeze see it as putting a stop to the force modernization process that they regard as the engine that drives the arms race. Some advocates go so far as to see the freeze as the first step toward not only a denuclearized world but one in which the role of force in international politics is no longer an acceptable mode of national conduct.

There are three principal objections to the freeze. The first is that important elements of it cannot be verified. This is a contentious issue on which knowledgeable people are divided. There is little doubt that full-range testing of ballistic missiles can be fully verified. One could adopt counting rules so that the number of warheads observed being tested on a missile would count as the number of deployed warheads for all missiles of this type. The locations of the production of fissionable material are also identifiable, although the output of these facilities may not be precisely determinable. But the numbers and locations of cruise missiles, especially sea-launched cruise missiles, cannot be verified with high confidence. Nor can weapon system component testing be easily detected. A second objection is that the freeze would "freeze" several disparities (especially ICBM hard-target kill capabilities) that favor the Soviet Union. And a third objection is that important nonnuclear capabilities (particularly antisubmarine warfare systems) would remain unconstrained, which could promote instability if the number of nuclear systems (including ballistic missile submarines) was frozen. Up to now widespread public support for the freeze has not been translatable into sufficient congressional support, nor has there been any alteration in the Reagan administration's opposition to the proposal.

Build Down

In the spring of 1984 the administration, following congressional pressure, introduced at the START (strategic arms reduction talks) negotiations in Geneva a "build-down" formula that was greeted coolly by the chief Soviet negotiator. The concept behind the build down, which has several complex variations, is that for each new nuclear warhead or delivery vehicle introduced into the arsenals of the superpowers at least twice as many existing systems would have to be discarded. Therefore, while force modernization would continue, the number of deployed forces would decline. The principal problems with this proposal are that (1) it would not constrain the mix of forces, so that the net result could be the introduction of more destabilizing weapons as the number of obsolescent weapons is reduced, and (2) it does not lend predictability to the arms competition, a major confidence-building goal of the arms control process. Political liberals have been particularly opposed to this concept on the grounds that it would be a stimulus to the arms race under the guise of arms control.

Midgetman

The President's Commission on Strategic Forces (the Scowcroft Commission) recommended in January 1983 that serious research and development commence on a single-warhead missile, now commonly known as Midgetman. The rationale behind this proposal is that missiles with MIRVs are the principal sources of instability because they enhance the cost-exchange ratio in favor of the attacker (that is, an attacker can use a small percentage of his MIRVed missiles to destroy a large percentage of the adversary's missiles). Deploying single-warhead missiles would be a step in the right direction. A world of only single-warhead missiles would be more stable because the attacker would be commiting mutual disarmament if an attack of *n* missiles would at best destroy *n* opposing missiles. The attractiveness of this proposal, therefore, hinges on the likelihood of our being able to move to a world without MIRVs. But the requirement to move to a deMIRVed world is extraordinarily demanding, unlikely to be achieved, and difficult to verify. If Midgetman proves merely to supplement existing systems, rather than replace them, it will serve no purpose of arms control.

No-First-Use Pledge

Besides proposals to constrain or reduce nuclear force deployments, there have been calls for the United States to pledge, as the Soviet Union has already done, not to be the first to use nuclear weapons. Advocates see this as a confidence-building measure and as a stimulus for the European allies to build up their conventional forces on the premise that they would not be able to rely on U.S. nuclear retaliation in the event of a Soviet conventional attack in central Europe. Opponents of the idea claim it would undermine alliance cohesion, that it would reduce uncertainty for Soviet military planners, and that there is no evidence to suggest that it would produce the desired buildup in European conventional forces. (This proposal is examined within the context of European security in chapter 7.)

Comprehensive Test Ban Treaty

There has long been support to conclude a comprehensive nuclear test ban treaty. The rationale behind it is that prohibiting testing would diminish confidence in the weapon systems, which would reduce the likelihood of their use (note, however, that the atomic bomb dropped at Hiroshima was of an untested design). Arguments against a treaty are that it (1) cannot be verified at low weapon yields; (2) would reduce stockpile reliability, which is important for deterrence and in case deterrence fails; and (3) would drive away qualified scientific and engineering personnel from working on nuclear weapon systems research and development. It is argued that this technical capability must be maintained to guard against an abrupt Soviet withdrawal from the treaty.

Antisatellite Weapons and the Military Uses of Space

Preventing the development of weapon systems that would be able to attack early warning and communication satellites would presumably be in the interest of both superpowers. A ban on antisatellite (ASAT) weapons, ground-based and space-based, together with a ban on the development of weapon systems of any kind in space, would be an important arms control achievement. (Note that the Outer Space Treaty now prohibits the deployment of nuclear weapons in space.) The Reagan administration has shown little interest in these proposals because it claims that they are not verifiable. Moreover, they cut against the president's strategic defense initiative, which has as its goal to render nuclear weapons "impotent and obsolete." Others see the United States as being able to gain a major technological advantage in space over the Soviet Union and are reluctant to negotiate it away. Differences between the superpowers over this issue must be narrowed if U.S.-Soviet arms control agreements are to have a future.

Conclusion

On balance it would seem that there are certain opportunities for arms control if the following conditions are met.

—There must be a clear presidential commitment to arms control and strong presidential leadership to guide the interagency negotiating process, the international negotiations, and the politics of ratification.

—Arms control must be coupled with force modernization and be sold as *part* of national strategy. It alone cannot carry the weight of being an alternative to national strategy.

—Simplicity is a highly desirable characteristic of any agreement. The more complex the document, the easier opponents can identify one element upon which to base their opposition. The greater the complexity, the more difficult it is to sell the agreement in domestic politics.

—Numbers of weapon systems or warheads tell us little about stability. What is crucial is the relative vulnerability of each side's strategic forces. The aim of arms control should be to reduce vulnerabilities, not to reduce numbers per se.

—Related to the objective of reduced vulnerability are the goals of transparency and warning. Any steps taken to increase the predictability of force deployments or to increase warning time to strengthen decision-making processes in crisis are highly desirable and should be thought of as part of arms control.

—Exclusionary agreements—the Outer Space Treaty, the Sea Bed Treaty, the Antarctic Treaty—have been negotiated in the past. It is easier to preclude weapon deployments from a region where they have never been then to regulate those that are already in place. To deny the militarization of outer space, therefore, has at least some encouraging precedents.

—We must be realistic about what arms control can achieve. Given the deep political, economic, and philosophical divisions between the superpowers, the arms race is more a symptom than a cause of the competition. At best, arms control agreements would constrain, but not transform, this competition.

If arms control is to play a constructive role in foreign and defense policy, it must be seen for what it is intended to be: threat control. Each side seeks to reduce the threats to its own society and, in military terms, to minimize the vulnerability of its forces. In the latter case, arms control is but one of several means—others include deception, mobility, and defenses—to satisfy this objective. Insofar as arms control can contain threats, especially threats against a country's capability to retaliate, it is a valuable diplomatic instrument that could help reduce the uncertainties of force planning. If arms control is to succeed, it must demonstrate through negotiated agreements that both sides have the political will to reach mutually satisfactory formulas that control the threats to them. If arms control achieves threat control, then all kinds of political payoffs are also within grasp. If, however, major threats continue unabated despite arms control negotiations and agreements, political opposition will eventually halt the process altogether. In short, for arms control to succeed and even continue to exist, it must control threats.

This conclusion has several implications.

—The United States ultimately found both the SALT I Interim Agreement and SALT II agreements wanting because they failed to control the principal Soviet threat to U.S. retaliatory forces, namely Soviet MIRVed ICBMs. Quantitative and even qualitative limitations established by negotiated agreement, no matter how comprehensive, cannot sustain political support if a principal challenge to U.S. forces remains uncontrolled.

—It is thus far more significant militarily and politically to control a single effective counterforce system than to reach agreed numerical ceilings on a wide range of forces.

—Threat control does not have to be symmetrical. There is nothing in principle to block a one-for-one asymmetric agreement as long as each side believes that the net vulnerability of its forces has been reduced.

—Confidence-building measures are also a form of threat control.

—Threat control need not be consummated solely through formal negotiated agreements. A country may also adopt unilateral measures to reduce threats.

—Given the dual-personality trait of the American national character, a certain commitment to the negotiating process may be politically necessary, but maintaining the process is far less important politically and militarily than controlling specific threats.

CHAPTER SEVEN

U.S. Allies Between Entrapment and Abandonment

The fault, dear Brutus, is not in our stars
But in ourselves. . . .

Shakespeare, *Julius Caesar*

ALLIANCES are like marriages. They have to be worked at continuously or they fall apart. The great industrial nations of the West remain linked through the North Atlantic Treaty, a multilateral security pact that sixteen nations have now signed.[1] The United States and Japan are linked by a bilateral cooperation and security treaty established in 1960, which stipulates that the United States will help Japan in case of armed aggression against it. The trilateral relationship involving North America, Western Europe, and Japan is also enmeshed in a complex network of bilateral and multilateral economic arrangements that give tangible meaning to the concept of interdependence. Yet after several decades of stormy but healthy marriages, by the early 1980s both the multilateral NATO security pact and the bilateral treaty with Japan were undergoing severe strain. Experienced observers characterized many of the differences as having the potential for doing permanent injury to both sets of relationships.

Nuclear weapons play only a part, though a crucial one, in each

1. The countries that signed the treaty in 1949 (the original members of NATO) are Belgium, Canada, Denmark, France, Great Britain, Iceland, Italy, Luxembourg, the Netherlands, Norway, Portugal, and the United States. Greece and Turkey signed in 1952, West Germany in 1955, and Spain in 1982. In 1966 France withdrew from the integrated military command structure but remains a member of NATO. The United States is also tied to Australia and New Zealand through the ANZUS pact and has security treaties with the Republic of Korea and other states, but these relations are not discussed here.

alliance system. Their roles can be accurately assessed only in the context of both sets of relationships. The central issue in each alliance system is that for different political, geographic, military, and psychological reasons, U.S. allies depend on American guarantees for their own security and are constantly trying to find some middle ground between entrapment or abandonment by the United States.[2]

For the allies, who have fully recovered from the ravages of World War II, routine support of American national security policies incurs opposition from domestic political forces whose purpose is to preserve allied independence from the United States. Moreover, each major U.S. ally has foreign policy interests often at odds with those of the United States. For an ally to adopt American positions often means supporting positions harmful to its own self-interest. Allies must therefore adopt positions independent from that of the United States, but not so independent as to lead to America's abandonment of the alliance.

This dilemma is particularly acute in relation to nuclear weapon policy, which is obviously central to the security of every member of the alliance and which is a politically charged issue within each country. In light of the constraints imposed on and by America's principal allies, the United States has maintained its guarantees to their security in part through the threatened use of nuclear weapons. The effectiveness of this policy of "extended deterrence" has long been judged in terms of the credibility of that threat.

Nuclear Forces for the European Theater

Historically, NATO rationalized its possession of fewer conventional forces than those of the Warsaw Pact by claiming that NATO could deter an attack because of its superiority in short-range nuclear systems and, ultimately, the nuclear umbrella provided by U.S. central strategic systems. This state of affairs allegedly convinced NATO planners that the West could control escalation of any potential conflict in central Europe. The policy of flexible response has been the explicit organizing concept of Western defense since the early 1960s, indicating the West's

2. The identification of the entrapment-abandonment dilemma was stipulated in the early 1960s. See Robert Endicott Osgood, *NATO: The Entangling Alliance* (University of Chicago Press, 1962).

ability to prevail in a European war without escalating beyond the necessary level.

The adoption of flexible response did not preclude attempts to improve the capabilities of NATO's conventional forces. But since 1967, when flexible response was formally incorporated into NATO's strategic planning, NATO's conventional defenses as well as its nuclear escalation capability have increasingly become the subject of criticism.

Warsaw Pact countries have long had greater numbers of conventional forces deployed on the central front than have the NATO countries. Moreover, in recent years the Soviet Union has deployed tactical aircraft and short-range nuclear weapons that can destroy NATO's battlefield nuclear weapons. After the United States relinquished its position of strategic nuclear superiority over the Soviet Union by the late 1960s, subsequent changes in the balance of intermediate-range (formerly called theater) nuclear forces began to produce real strains in the alliance.

The deployment of Soviet SS-20 IRBMs—mobile, high-accuracy systems with MIRVs—in the western military districts of the Soviet Union starting in 1976 progressively undermined the confidence of Western defense planners in NATO's capability to escalate to nuclear weapon use to counter a conventional Warsaw Pact attack. The IRBMs and conventional Backfire bombers, short-range Warsaw Pact systems and nuclear artillery, and high-performance conventional forces transformed the military situation, or at least the perception of the military situation. All NATO intermediate-range nuclear systems were thought to be increasingly vulnerable to attack.[3] This realization came at a time when growing U.S. ICBM vulnerability and trends in the Soviet-American strategic balance were creating doubts among Europeans about the credibility of the American nuclear guarantee.

Some observers concluded that control of nuclear escalation had passed to the Soviet Union. They believed that in the event of a conventional Warsaw Pact attack on West Germany, the firepower and numerical superiority of Pact units would force NATO troops to retreat rapidly, and NATO officials would have to consider use of nuclear

3. It has been claimed that the SS-20s contributed only marginal additional military capability because the older SS-4s and SS-5s could already destroy NATO nuclear systems. See Stephen M. Meyer, *Soviet Theatre Nuclear Forces,* part 2: *Capabilities and Implications,* Adelphi Paper 188 (London: International Institute for Strategic Studies, 1984), p. 26. If correct, this was not the generally accepted view in the West prior to the SS-20 deployment.

weapons to stem the communist advance. The Soviets, with an intimate knowledge of NATO decisionmaking, would then strike preemptively (perhaps needing to use only high-accuracy conventional forces) at all NATO nuclear systems, including quick-reaction aircraft, nuclear artillery, Pershing IRBMs, and all other theater-based forces. The United States, it was argued, would fail to respond for fear of initiating a strategic exchange that would destroy the American homeland. So it was the combination of changes in the European-theater military balance coupled with the shift in the strategic balance that caused concern in European capitals and then in Washington. This concern finally resulted in the December 1979 NATO decision calling for deployment of 108 Pershing II IRBMs and 464 GLCMs starting in late 1983, contingent on the commitment to seek a negotiated arms control agreement with the Soviet Union on intermediate-range nuclear systems. (Initial deployments did indeed commence in December 1983 after no significant progress had been made at the U.S.-Soviet intermediate nuclear force negotiations in Geneva.)

The web of political complexities in which these intermediate nuclear force (INF) deployments have become entangled is illustrated by the siting problem for INF missiles. The West Germans were quite naturally concerned about the threat to NATO's nuclear systems. Consequently, then Chancellor Helmut Schmidt and the conservative members of the Social Democratic party sought modernization, but because they did not want Germany alone to bear the brunt of INF modernization, they demanded "burden sharing"—that is, missile bases in other countries as well as Germany. Many Germans thought Britain was too removed, at least psychologically, from the European continent. Thus, although improvement in British nuclear forces was welcome, it did not satisfy German concerns. Norway and Denmark were off limits since both had policies prohibiting the stationing of nuclear forces on their territory in peacetime. Greece and particularly Turkey made good sense militarily because of their geographic position, but strained U.S. bilateral relations with each country and political instability in both ruled them out. France was not a candidate since it was no longer within NATO's integrated military command structure. That left Belgium, the Netherlands, and Italy as the three candidates for burden sharing. The Italians made the commitment to deploy missiles, contingent of course on whether the government in power at the deployment date would honor the original pledge. The Belgians would not commit themselves to deploy missiles without Dutch acquiescence, and the Dutch seemed to tie their support to the ratification of the SALT II treaty, which never materialized.

Since NATO's 1979 decision to deploy IRBMs and cruise missiles, several analyses have questioned the wisdom of such deployments on three fronts: the nonnuclear military balance; the appropriate doctrinal posture for NATO; and the organizational arrangements of the alliance.

Careful examination of how Warsaw Pact countries would conduct a blitzkrieg attack on the Central Front suggests that they could not win quickly (assuming neither side used nuclear weapons) and that NATO would have time to marshal its forces for an effective counterattack.[4] Subsequent studies have noted that with perhaps a 3 or 4 percent real increase in NATO defense expenditures for the next six to ten years, NATO could acquire the additional conventional and nuclear forces needed to mount a credible conventional defense of Western Europe.[5]

In two celebrated articles experts have supported the adoption of a NATO pledge of no first use of nuclear weapons on the grounds that (1) NATO has no current plan specifying how to use its nuclear weapons first, and it is doubtful that any sensible plan could be devised, and (2) the adoption of such a pledge would stimulate European countries to increase their defense expenditures and to procure the necessary conventional forces to acquire a credible nonnuclear deterrent.[6] Moreover,

4. John J. Mearsheimer, "Why the Soviets Can't Win Quickly in Central Europe," *International Security,* vol. 7 (Summer 1982), pp. 3–39. Mearsheimer's analysis, however, does not take into account the role of conventional aircraft or NATO's potential difficulties in responding in a timely fashion to the threat of a Warsaw Pact attack.

5. For one interesting proposal that accords with thinking at SHAPE (Supreme Headquarters Allied Powers Europe), see "Do You Sincerely Want to Be Non-Nuclear?" *Economist,* vol. 284 (July 31–August 6, 1982), pp. 30–32. See also European Security Study, *Strengthening Conventional Deterrence in Europe: Proposals for the 1980s* (St. Martin's Press, 1983); and William W. Kaufmann, "Nonnuclear Deterrence," in John D. Steinbruner and Leon V. Sigal, eds., *Alliance Security: NATO and the No-First-Use Question* (Brookings Institution, 1983), pp. 43–90.

6. See McGeorge Bundy, George F. Kennan, Robert S. McNamara, and Gerard Smith, "Nuclear Weapons and the Atlantic Alliance," *Foreign Affairs,* vol. 60 (Spring 1982), pp. 753–68; and Robert S. McNamara, "The Military Role of Nuclear Weapons," *Foreign Affairs,* vol. 62 (Fall 1983), pp. 59–80. General Bernard W. Rogers, Supreme Allied Commander Europe, has advocated a declaratory policy of no early first use in "The Atlantic Alliance: Prescriptions for a Difficult Decade," *Foreign Affairs,* vol. 60 (Summer 1982), pp. 1145–56. McNamara, as noted in chapter 5, claims to have urged both Presidents Kennedy and Johnson not to use nuclear weapons first despite the doctrinal pledge of flexible response. Moreover, William Kaufmann has now concluded that "NATO does not have and may never have had a particularly credible nuclear deterrent." See Kaufmann, "Nuclear Deterrence in Central Europe," in Steinbruner

the case has been made not only that NATO should strengthen its conventional forces to provide "reassurance" as well as deterrence, but also that Europe should take on more of the defense burden and create a European defense policy within the NATO framework.[7]

Strengthening NATO's conventional forces is a measure that has virtually unanimous support on both sides of the Atlantic. Moreover, it is probably true that none of the five categories of nuclear weapon use in the European theater—demonstration (to show resolve), limited defensive use (to turn the tide in local battlefield situations), restricted battle area use, extended battle area use, and theaterwide use—have been well thought through or could be before a conflict erupted. There are two especially contentious issues, however. The first is whether the emerging technologies reflected in NATO conventional force deployments could markedly strengthen the conventional deterrent. The second is whether the adoption of a no-first-use pledge would have the desired political and military effects.

Emerging Technologies and the European Balance

Within the next decade several emerging technologies are expected to create new classes of weapon systems. One class consists of precision-guided munitions (PGMs)—that is, munitions guided throughout their flight with a probability greater than 0.5 of directly hitting their targets at full range. The high accuracy of these systems comes from incorporating three types of precision-guidance technologies: seeker guidance, which uses, for example, laser or infrared devices, in which the weapon system "homes in" on the target; precision positioning guidance, which uses, for example, the yet-to-be-built NAVSTAR global positioning system, in which the weapon system receives signals from a synchro-

---

and Sigal, eds., *Alliance Security,* p. 38. More recently, to augment the policy of no early first use, General Bernard W. Rogers has pushed forward NATO's adoption of a revised tactical doctrine, known as follow-on force attack (FOFA), that calls for counterattacking an invading Warsaw Pact force far behind the front lines by using aircraft, conventionally armed missiles, and remotely deliverable mines. See Richard Bernstein, "General Defends NATO's New Tactical Doctrine," *New York Times,* November 14, 1984.

7. See Michael Howard, "Reassurance and Deterrence: Western Defense in the 1980s," *Foreign Affairs,* vol. 61 (Winter 1982–83), pp. 309–24; and Hedley Bull, "European Self-Reliance and the Reform of NATO," *Foreign Affairs*, vol. 61 (Spring 1983), pp. 874–92.

nized transmitter that allows it to correct its accumulated errors in flight; and correlation guidance, which uses, for example, a terrain contour-matching (TERCOM) technique in which the weapon system carries a computer that performs map-matching computations to compare the system's actual position to a preprogrammed position and makes the necessary in-flight corrections.[8]

A second class of weapon systems relies on remote guidance and control techniques, frequently television guidance. These remote-piloted vehicles (RPVs) are pilotless systems controlled from a distant location by an operator who has the same information he would have if he were on board. The RPV can be recalled before reaching its target and is capable of returning to base after completing its mission. A third class of weapon systems consists of a variety of improved munitions tailored to specific targets. Examples include scatter mines, fuel air explosives, and shaped charges.

Advances in high-frequency electromagnetic transmitters and receivers and in microelectronic integrated circuitry are leading to major improvements both in command, control, and communication systems and in target identification and acquisition capabilities. When coupled with enhanced propulsion efficiency on the one hand and improved warhead yield-to-weight ratios on the other, these advances will result in a family of weapon systems with high accuracy, long range, and controlled explosive power.

With respect to these technologies, three propositions deserve attention. The first proposition is that weapon technologies of precision and control will favor the defending NATO forces rather than the attacking armies of the Warsaw Pact.

It can be argued that the deployment of PGMs, RPVs, and target-tailored munitions would enhance NATO defense because they would greatly increase the vulnerability of concentrated attacking forces. They would be particularly effective in blunting the advances of Warsaw Pact tank armies. They would increase the lethality of NATO's ground-based antiaircraft systems. They would improve the ability of small units to defend NATO territory. And they would increase the vulnerability of logistical facilities in rear-echelon areas in Eastern Europe that support the attacking forces. These weapons, it can be argued, are better suited to a defender who lies concealed in familiar terrain than to an attacker

8. American long-range cruise missiles, whether air-launched, sea-launched, or ground-launched, may be considered PGMs with correlation guidance.

who must reveal himself when moving into unfamiliar territory.[9] There will exist, moreover, a synergistic effect among these weapon systems (for example, scatter mines will necessarily slow advancing tanks, in turn making them more vulnerable to PGM attack). In addition, cost-exchange ratios will strongly favor NATO, with weapons costing less than $50,000 aimed at major Warsaw Pact systems that cost $500,000 or more.

However, there are several reasons to be skeptical of this proposition. First, it is too broad to be useful. To evaluate the effectiveness of these new weapons in the European theater would require knowing specifics about opposing forces as well as climatic and terrain conditions. In the event of war, so-called offensive and defensive weapon systems will be used by both attacking Warsaw Pact forces and defending NATO forces, and it is therefore misleading to view these new technologies as primarily beneficial to the defender.

Second, there are operational limitations. NATO systems may not perform as well in combat as under test conditions. In the fog of war sophistication can be a drawback rather than an advantage. Most PGMs, moreover, are weather-constrained: they operate best in clear daylight and do not function as well at night and in foggy or cloudy environments. In addition, extraordinary command, control, and communication needs must be met for these sophisticated weapons to function effectively—needs that possibly cannot be satisfied because of the multinational and multilingual character of the NATO force and the sheer numbers of systems expected to be in operation by the end of the 1980s.

Third, the Soviets are fully aware of the potential effectiveness of PGMs, RPVs, and improved munitions and are perfecting countermeasures and adapting both strategies and tactics to nullify these systems on the battlefield. Besides taking advantage of weather conditions least favorable to the emerging technologies, the Soviets would most likely use jamming techniques, camouflage, and smoke to degrade the performance of these weapon systems. Improvements incorporated into Soviet forces over the last several years indicate a determination to overcome any advantage NATO might gain through acquiring new sophisticated weaponry. Indeed, it has been argued that the Soviets are now reducing their emphasis on the tank and moving to a combined-arms approach that emphasizes both motorized infantry and artillery, using mechanized

9. At sea the new weapons would again favor the defender because the vulnerability of large surface ships would make amphibious assaults more difficult.

infantry combat vehicles. The Soviet combined-arms concept is based on suppression, maneuver, defense, and combat support. The suppression function involves using artillery, mortars, multiple-rocket launchers, surface-to-surface missiles, close air support, and helicopter gunships to destroy NATO's antiarmor defenses, artillery, and mortars as well as its nuclear battlefield command posts and fire-control centers. Soviet strategy calls for advancing rapidly to overwhelm NATO defenses, relying on surprise and massive firepower from self-propelled armored artillery.

Fourth, the proposition is misleading. Weapons of precision and control alone are not likely to have a strong effect on NATO defenses because the weapons tend to conceal rather than resolve many fundamental alliance problems. Foremost among these problems is the inferiority complex that grips Western European nations, nations that are militarily inferior to Warsaw Pact countries more because they think of themselves as small powers than because they lack economic or technological resources. Operational problems flowing from this sense of inferiority are compounded by the inability of alliance members to coordinate more effectively among themselves. NATO force deployments are not well designed to meet the principal Soviet threat on the North German Plain. Neither combat nor reserve units are maintained at adequate levels of readiness. NATO's weapon and logistical systems tend not to be as interdependent as they should be. These problems, coupled with shortages in weapon stocks and difficulties in the manning of emergency defense positions, far outweigh the more publicized and probably insoluble problem of lack of standardized weapons. Without correcting these shortcomings, NATO's adoption of weapons of precision and control would have little effect on its defense capabilities.

The fifth problem is that of urban sprawl. The development of large urban areas in northeast Germany over the last several years and the projected growth of these areas during the next decade will hamper NATO's use of PGMs. Because West Germany continues to build more roads and because its rural areas are heavily wooded, Warsaw Pact countries would probably attack first in an urban area, forcing the NATO defenders to choose whether to destroy precisely what they would be trying to preserve. Although striking urban areas first would slow the Warsaw Pact rate of advance, it would have the advantage of complicating NATO's use of antitank weapons, which are least effective at very short ranges against mobile targets.

Sixth, there is organizational and political resistance to emerging weapon technologies within NATO itself. For the new weapons to be most effective, they would have to be procured in large numbers by a NATO that would be prepared to adjust its tactics and organizational structure to the capabilities these weapons would provide. NATO's past behavior indicates that it probably would not be that receptive to making such adjustments. On economic grounds, the fact that much of this advanced weaponry will be manufactured in the United States for at least the next five to ten years is generating resistance among European industrial and labor groups. These groups quite naturally want to promote procurement decisions within their own nations, which would benefit them economically. European governments are therefore under considerable pressure to steer away from procuring PGMs, RPVs, and improved munitions unless such procurements translate directly into jobs and profits at home.

Receptivity to new weapon technologies is also lagging because of service interests. The military communities of NATO countries are attempting to incorporate these technologies into existing doctrines, missions, and procedures incrementally, even though nonincremental adaptation is what is necessary. The reason: the procurement of PGMs, RPVs, and improved munitions is not welcomed because these technologies threaten to compete with those systems—main battle tanks and modern fighter aircraft—that are the essence of the military's mission.

The seventh objection to the proposition that emerging weapon technologies will favor NATO defense is that these technologies may cost more than originally expected. Although the current generation of PGMs is relatively affordable, subsequent generations may not be so cheap. The systems will need to incorporate increasingly sophisticated electronic equipment to negate Soviet countermeasures and be mounted on mobile and highly maneuverable launch platforms to reduce their vulnerability—two expensive requirements. In addition, although procurement costs for these new weapons may remain low in relation to the worth of the targets they may defend, replacement and life-cycle costs will substantially reduce their cost-benefit advantage.

A second proposition that has been advanced regarding the emerging weapon technologies is that they will stabilize the military balance in central Europe by raising the nuclear threshold and adding to the deterrent value of NATO forces. The proposition is defended on the grounds that weapons of precision and control will have the unprecedented capability of hitting and destroying the targets at which they are aimed while causing a minimum of secondary damage. When fully incorporated into doctrine and force posture, this capability should significantly enhance the alliance's ability to actually implement the strategy of flexible response, which has long been accepted policy. NATO's ability to withstand Warsaw Pact advances without resorting to nuclear weapons will be increased, thereby raising the nuclear threshold. These new weapons will consequently provide Western political leaders readily usable options that eliminate the need to escalate from conventional to nuclear weapons. Such options would enhance the confidence of NATO leaders and impress the Soviets with the credibility of Western defenses.

Working against this proposition, however, is the notion that deterrence in central Europe is linked to the fear that any Warsaw Pact conventional attack will inevitably lead to the use of tactical nuclear weapons and, eventually, to a Soviet-American nuclear exchange on the Soviet and U.S. homelands. The inability of the attacker to limit the conflict to conventional weapons would itself be a deterrent. If NATO relies more heavily and explicitly on conventional weapons, no matter how sophisticated, it will appear to the Soviets that NATO is reluctant to use tactical nuclear weapons. War therefore becomes more thinkable and thus more probable. It could be argued that the minimum secondary damage the new conventional weapons would cause reduces rather than increases risks to the enemy should he attack first. Critics of this proposition therefore claim that for deterring Soviet aggression it is better to maintain the distinction between conventional and nuclear weapons, to emphasize the possible use by NATO of nuclear weapons in the event of a Warsaw Pact attack, and to underscore the unpredictability and horror of nuclear war.

A third proposition is that emerging weapon technologies will channel arms control efforts toward the reduction of valuable offensive systems and that in time the emphasis in central Europe for both NATO and Warsaw Pact forces may be on developing defensive capabilities. It can be argued that the demonstrated effectiveness of weapons of precision and control will stabilize the European military balance, thus providing the impetus to constrain the development of the large, complex, and costly weapon systems—particularly tanks and manned fighter aircraft—most vulnerable to attack by these new weapons. Given the escalating costs of these major systems, their growing vulnerability, and the defender's continued advantage of benefits in relation to cost, war strategies could increasingly emphasize military rather than civilian targets. The principal arms competition would then center on defensive capabilities.

However, the emerging weapon technologies will exacerbate the inequalities between NATO and Warsaw Pact forces. Consequently, arms control will be less rather than more likely to flourish. Moreover, the proliferation of gray-area systems is making obsolete the distinction between strategic and tactical weapon systems, a distinction that was always artificial for operational purposes but that was convenient for conceptual clarity and the arms control negotiation process.

It appears overall that the introduction of emerging technologies is complicating the assessment of the military balance in central Europe, which is presumably a welcome development from the perspective of deterrence. But the conclusion that either side's acquisition of the new weapons would strengthen the deterrent against conventional war cannot be easily substantiated.

Centripetal and Centrifugal Forces in NATO

For NATO to continue to fulfill its mission of defending Western Europe, it must retain political cohesion. Since the Federal Republic of Germany entered NATO in the mid-1950s, the relationship between Germany and the United States has been at the heart of the NATO alliance. In the following decades common estimates of German and American mutual security interests and their close economic and cultural ties have bound the two countries together. The realities of both German domestic politics and European politics are that Germany cannot shoulder too much of the defense burden for NATO. In the face of the ever-present Soviet threat, the American military presence in Western Europe and the American nuclear guarantee provide cohesiveness that holds the alliance in place; these are the centripetal forces.

But if the United States was seen as backing away from its security guarantee by leaving more of a role to the Germans, the other NATO countries would feel abandoned, which would be disastrous for the alliance. Many non-German Europeans and Germans themselves fear that a weakening of the American commitment to NATO would lead to the emergence of the Federal Republic as the military and political leader of Europe, a condition that France and the Soviet Union would find intolerable. Within Germany, the desire for reunification of the country remains latent; the absence of American support could stimulate the country to reach a rapprochement with the Soviet Union, which would do irreparable harm to American security interests.[10]

How would an American no-first-use pledge affect these concerns? European governments could have one of four responses to such a pledge: no reaction; a shift toward conventional defenses; a shift toward nuclear proliferation; or a shift toward the Soviet Union, neutralism, or pacificism. It is not clear which response is likely or if each government would respond differently. However, retaining the option of first use, which does not *require* first use, seems preferable to a pledge in peacetime that would probably be meaningless in wartime.

Nuclear weapons are not a substitute for conventional strength in Europe, but conventional forces alone cannot provide the political and psychological symbols required to maintain a cohesive alliance system.

Japanese Defense

Although Japan is not now a major factor in American nuclear weapon policy, it has the potential to become a strong military power with a force posture intimately tied to the U.S.-Soviet nuclear balance. Moreover, the Japanese have the same fear of being caught between entrapment and abandonment that Europeans are struggling with.

Since Japan began to recover economically after World War II, American defense planners have thought of that island nation as an element in the strategy of U.S. containment of communist expansion in the western Pacific. This was not a new role for Japan. The Japanese had been trying to contain the expansion of Russian influence since the late nineteenth century; their efforts culminated in their spectacular victory in the Russo-Japanese war of 1904–05. After World War II,

10. Note that West Germany is the largest Western trading partner for both the Soviet Union and East Germany, that roughly three in ten West Germans can identify a relative in the East, and that approximately 80 percent of the East German population are able to watch West German television nightly. For West Germany, far more than for any other state in the Western alliance, improved East-West relations has had a tangible and lasting effect.

American defense planners saw Japan as a launching platform near the Asian mainland for American armed forces intent on preventing the spread of communism. They also saw Japan as the forward line of defense for American assets in the Pacific, which stretched from Okinawa and Guam to the Hawaiian Islands and on to the Pacific coast of the United States.

After the American defeat in Vietnam and the collapse of détente with the Soviet Union, anti-Sovietism increasingly replaced anticommunism as the organizing principle of American foreign policy. Central to this principle has been the formation of an implicit alliance among Western Europe, Japan, China, and the United States to resist Soviet expansion politically, economically, and militarily. The Reagan administration has sought to foster this process wherever feasible.

In this newly emerging geopolitical alliance, Japan plays several crucial roles. The most important, perhaps, is its symbolic presence as a politically stable democracy that has achieved the status of an economic superpower. Japan is one of the few democracies in Asia, it has built the second most productive economy in the noncommunist world while remaining closely allied with the United States, and it functions peacefully in the commonwealth of nations. In the event of an American-Soviet military confrontation, Japan could play four additional roles.

First, U.S. bases in Japan could provide critical air and naval support for American military forces engaged in renewed hostilities on the Korean peninsula. Second, Japanese maritime and ground forces could deny the Soviet navy access to the Pacific Ocean by blocking the Korean, Tsugaru, and Soya straits. Third, Japanese naval forces could protect the sea lines of communication near Japan and the western Pacific if the U.S. Seventh Fleet became engaged in a conflict in the Persian Gulf or Indian Ocean. Fourth, Japanese military forces could prevent the Soviet Union from opening a Far Eastern front in the event of a U.S.-Soviet conflict in Europe or southwestern Asia.

The Soviet Union, in the event of hostilities or the serious threat of hostilities with the United States, would like to be able to coerce Japan into submitting to Soviet demands or surrendering unconditionally; to destroy Japanese and American military bases to prevent their forces from interdicting Soviet sea lines of communication; and to gain control of the three straits to ensure free passage of the Soviet navy into the western Pacific. Consequently, for Japan to contribute most effectively to U.S. global strategy against the Soviet Union, it must be able to deny the Soviets the capacity to achieve these objectives. This Japanese capability would complicate Soviet defense decisionmaking and cause the Soviet leaders to think twice before initiating military actions in areas of vital interest to the Western alliance.

There are several missions, however, that Japan is not being asked to perform. The United States would not want Japan to provide combat forces to fight in Korea, in the Philippines, or anywhere else in East Asia. The United States does not wish Japan to acquire the surface naval forces necessary to protect its own sea lines of communication in the area from the Persian Gulf to the Sea of Japan. Indeed, U.S. strategy does not call for any truly offensive Japanese military capability, whether on land, at sea, or in the air. Moreover, the United States has no contingency plans specifying the need for Japan to acquire an independent nuclear deterrent. In short, in the event of hostilities between the superpowers, the United States would expect Japan to play a limited role, emphasizing the defense of the Japanese homeland and the mission of denying the Soviet navy access to the western Pacific.

In the last decade much has happened to shake Japanese confidence in the United States: the precipitous American rapprochement with China, symbolized by President Nixon's spectacular visit there in 1972; the soybean embargo of 1973; the humiliating U.S. defeat in Vietnam, culminating in the communist victory in 1975; and President Carter's decision in 1977 to conduct a phased withdrawal of all U.S. ground troops from South Korea, followed by the reversal of this decision a year later. These shocks have reduced the confidence of the Japanese public in the credibility of U.S. security guarantees to Japan. The shift in the U.S.-Soviet strategic nuclear balance, the introduction of Soviet SS-20 intermediate-range ballistic missiles and Backfire bombers into the Pacific theater, and the growth of the Soviet naval presence in northeast Asia during a period of American naval retrenchment in that region have reduced U.S. credibility even further.

Yet despite these shocks and adverse trends, the Japanese public is not greatly concerned about a Soviet military threat. Although the Soviet Union is widely disliked and distrusted in Japan, most Japanese do not believe it poses a direct military threat to their country, nor do they take seriously the idea that the Soviet Union would try to cut off their supplies of oil from the Middle East.

While the United States has traditionally placed Japan within the context of U.S. global strategy, Japan has mostly concentrated on local defense and the role of the United States in preserving the national security of Japan. This contrast between the American global and regional perspective on the one hand and the Japanese local perspective on the other is an important distinction between the two nations concerning security.

Japanese public opinion polls reveal not only a deep distrust of the Soviet Union but also strong support for the Japanese self-defense forces and for the maintenance of Japan's security treaty with the United States. However, powerful political, psychological, and economic forces are constraining Japanese support for greater military power. Many feel that, judging from the experience of World War II, a high Japanese political-military profile in international politics would lead to the militarization of Japanese society, with ultimately disastrous effects on the nation. The sobering fact that Japan is the only country to have experienced nuclear annihilation is a permanent scar on the Japanese psyche. In addition, Japan in the postwar era has experienced extraordinary prosperity by concentrating on economic affairs and leaving "high politics" to the United States. It is no easy matter to alter a formula that has produced such marked success.

Japanese are also constrained by their concerns about the reactions of neighboring states to the strengthening of their military forces. They are well aware that Japanese militarism in the 1930s deeply alienated Koreans and southeast Asians, so they are reluctant to adopt a more visible military posture in the face of a questionable Soviet threat, a posture that might contribute little to peace and stability in East Asia.[11]

In general, then, most Japanese could be classified as minimalists on security matters and are largely content to react to American pressures and adopt only those measures that are essential to alleviate such pressures.

There are other perspectives, however. The gradualists include most members of the Liberal Democratic party, officials in the Foreign Ministry and the Japanese Defense Agency, and students of strategic studies in the Japanese Defense Academy, the National Defense College, and some think tanks. They support a more assertive Japanese defense posture designed to defend the Japanese islands against a Soviet conventional attack. They argue that Japan should do more than merely

11. The contrast between popular and elite views of Japanese defense policy is examined in Mike M. Mochizuki, "Japan's Search for Strategy," *International Security*, vol. 8 (Winter 1983–84), pp. 152–79.

respond to American pressure, not only because showing more initiative makes good sense from the perspective of a sound defense policy, but also because such assertiveness would ease strains in U.S.-Japanese economic relations.

At the extremes are the neutralists and the Gaullists. The neutralists, whose principles are embodied in the official platform of the Japanese Socialist party, support unarmed neutrality and the severance of alliance relations with the United States. Public support of this position is declining and will probably continue to do so in the near future. The Gaullists, however, are more significant. This group includes members of the conservative wing of the Liberal Democratic party, civilians in the Japanese Defense Agency, uniformed officers in the self-defense forces, and some defense intellectuals and businessmen. They think Japan should loosen or sever security ties with the United States and adopt an independent military posture that would include a sea-based nuclear deterrent force modeled along the lines chosen by General Charles de Gaulle in France in the 1950s and early 1960s. The withdrawal of all U.S. forces from Japan and the deployment of a large surface navy with aircraft carriers are elements of the Gaullist approach. The Gaullists argue that the United States can no longer credibly guarantee Japanese security and that it is time that the Japanese acquire a military capability commensurate with their economic strength. If American pressures, stemming from economic or security grievances, lead to a collapse of the U.S.-Japan security relationship, it will probably be the Gaullists—capitalizing on deep feelings of Japanese nationalism—who will dominate Japanese defense policy rather than the minimalists, gradualists, or neutralists.

Except for the neutralists, all groups now support the various measures of defense cooperation that the Japanese and American military establishments have undertaken in recent years. These include joint exercises, exchange of intelligence information, mutual logistical support, coordination of the readiness stages of U.S. and Japanese forces, and division of defense responsibilities by function and geography. The broadening and deepening of this cooperation would enhance Japanese security and reduce strains in the defense relationship with the United States.

Japan's military might is not negligible. Japan has the eighth largest defense budget in the world and in recent years has purchased from the United States or agreed to produce under license major sophisticated weapon systems including F-15 combat aircraft, E-2C reconnaissance aircraft, and P-3C antisubmarine warfare aircraft. Moreover, the Japanese have greatly increased their share of the costs of maintaining U.S. forces in Japan.

Nonetheless, the scope and quality of the Japanese defense effort is deficient in two crucial respects: it could not counter or deter possible Soviet actions and it is not strong enough to appease increasing American resentment that the United States is providing a "free ride" in defending Japanese security while Japan devotes its energies to strengthening its commercial economic competitiveness.

The Japanese know that misplaced American pressure could stimulate a polarization of views on Japanese security. This could result in a stronger position for the minimalists or the ascendancy of the Gaullists. Neither development would serve the interests of American or Japanese security.

A strong alliance with NATO on the one hand and with Japan on the other is central to U.S. security interests. The German or the Japanese economy at the service of the Soviet military establishment would be disastrous to the United States. Consequently, maintaining cohesive alliances is among the most vital objectives of American foreign policy.

In the next two decades Germany and Japan in particular will be defining roles for themselves in ways they have not done before. These defeated powers have recovered economically but not psychologically: they are still afraid of their past and of themselves. But by the end of the twentieth century this fear will have dissipated with the rise of a new generation. Maintaining a policy that safeguards U.S. allies from either entrapment or abandonment is, therefore, in the interests not only of the allies but of the United States. Continuous marginal improvement in U.S. and allied conventional forces, the maintenance of American nuclear guarantees, and an acute sensitivity in Washington to domestic political and economic trends within the domestic societies of the alliance are collectively the most appropriate recipe for maintaining alliance cohesion and the credibility of extended deterrence.

CHAPTER EIGHT

Nuclear Proliferation and Nuclear War

If you can look into the seeds of time,
And say which grain will grow and which will not,
Speak then to me. . . .

Shakespeare, *Macbeth*

A FINAL dilemma in the age of vulnerability concerns the acquisition of nuclear weapons by new nations and the effect of this process on the likelihood of nuclear war. The United States has taken the lead since World War II in diplomatic efforts to limit the spread of nuclear weapons to new nations. The U.S. initiatives have included the Acheson-Lilienthal Report, the Baruch Plan, the Nuclear Nonproliferation Treaty, the nonproliferation policies of the Carter administration to limit the exploitation of nuclear-energy facilities and materials for nuclear weapons development, as well as the Limited Test Ban Treaty, the Seabed Treaty, the Outer Space Treaty, and other bilateral and multilateral arms control measures. But despite these efforts, nuclear proliferation has continued, although at a much slower pace than most analysts expected.

The problem of nuclear proliferation is complex and intimately related to questions of national security, prestige, energy policy, and domestic politics that vary considerably among nations. It has long been argued, particularly by citizens of less developed countries, that there is a strong connection between vertical proliferation—the continued growth of the nuclear arsenals of the United States and the Soviet Union—and horizontal proliferation—the spread of nuclear weapons to nonnuclear states. It is argued that the Nuclear Nonproliferation Treaty is inherently discriminatory because it preserves the right of the nations that already have nuclear weapons to retain their nuclear arsenals but denies the right of countries without nuclear weapons to acquire them. Moreover, the treaty specifies that the nuclear powers party to it agree to reduce substantially their nuclear forces—a commitment that the United States and the Soviet Union have both failed to honor. Some spokesmen for countries without nuclear weapons argue that unless this commitment is met, the nuclear powers have no legal or moral justification for seeking to prevent the spread of nuclear weapons to other countries.

These countries have at least three incentives to acquire nuclear weapons. Some see nuclear weapons as the ultimate guarantee of their national sovereignty. For these nations, such as Israel and South Africa, there is probably little or no relation between their desire to acquire nuclear weapons and the behavior of the Soviet Union or the United States with respect to strategic arms control. Other countries, such as Argentina and Brazil, see nuclear weapons as a symbol of political prestige, both foreign and domestic. Within these countries there is likely to be vigorous debate about whether the nuclear threshold should be crossed.

Proponents of nuclear weapon acquisition may well use the horizontal-vertical argument as a rationale for supporting their position. Countries unsure about acquiring nuclear weapons, however, may sincerely believe that the nonproliferation stand is discriminatory. These countries may decide against or in favor of nuclear weapons on the basis of whether the Soviet-American nuclear appetite is curbed. Still other states might find nuclear weapons useful both as a response to an incipient nuclear rival and for domestic political purposes. Pakistan, influenced by the Indian nuclear program, and Iraq, concerned about the nuclear weapon potential of Israel, fall into this category. Soviet and American nuclear weapon postures probably do not influence the decisions of these countries very much.

One must recognize, however, that there are also powerful disincentives to acquiring nuclear weapons that have surely had an inhibiting effect on the nuclear proliferation process. These include cost (tens of billions of dollars for the research, testing, development, and deployment of a modest stockpile of weapons and their associated delivery vehicles); the demanding technical requirements of weapon design and fabrication; the operational problems of deploying a secure and survivable force; and the formidable domestic opposition in some countries to the acquisition of a military capability that they suspect would provoke their neighbors and the superpowers and produce a net decrease in their security. One or more of these reasons has dissuaded Italy, Sweden, Japan, and others from acquiring nuclear weapons.

In the considerable volume of work done in the last two decades on the nuclear proliferation problem, it is surprising that so little attention has been given to the relations between the spread of nuclear weapons and the security policies of the United States. Most of the analyses that have addressed questions of American policy have focused solely on the nonproliferation issue. What is less well understood, however, is how American foreign and military policies and major U.S. defense programs affect and are affected by the acquisition of nuclear weapons by other countries. What these effects have been in the past is a clue to what they might be in the future. The risks of nuclear weapon use in countries that have recently acquired them are great enough to directly affect American security policy.

The Past

In the four decades since the United States detonated two atomic bombs over Japan, four countries have stockpiled large numbers of nuclear weapons and acquired sophisticated means for their delivery, one state has exploded a single nuclear device and has evidently not acquired more weapons, and one state has become able to detonate one or more nuclear weapons on short notice. The following discussion describes how American security policy has been affected in the first instance by Soviet, British, French, and Chinese nuclear weapon programs, in the second instance by the Indian nuclear detonation, and in the third by Israel's ambiguous nuclear weapon status.

The Soviet Union

America's adversarial relationship with the Soviet Union after World War II was evident before the Soviets acquired nuclear weapons. The famous "long telegram" issued in February 1946 by George Kennan, then chargé d'affaires at the U.S. embassy in Moscow, is widely credited with having shaped thinking in Washington about Soviet behavior. Kennan argued that Soviet hostility toward the West enabled Russian leaders to feel secure internally and to justify their totalitarian rule. Kennan concluded: "We have here a political force committed fanatically to the belief that with the U.S. there can be no permanent *modus vivendi*, that it is desirable and necessary that the internal harmony of our society be disrupted, our traditional way of life be destroyed, the international authority of our state be broken, if Soviet power is to be secure."[1]

Subsequently, of course, Kennan articulated his views publicly in the famous *Foreign Affairs* article that introduced containment as the organizing concept of American policy toward the Soviet Union. But the article did not refer to the prospect of the Soviet Union acquiring nuclear weapons. Containment, as seen by President Truman, Secretary of State Dean Acheson, and other high-ranking American officials, was surely predicated on other considerations: the threatening nature of Soviet communist ideology; the consolidation of Soviet political control in Eastern Europe; the potential threat to Western Europe posed by Soviet conventional forces; the subversive techniques of the international communist movement controlled by Moscow; and the intransigence of Soviet negotiating behavior on a wide range of postwar issues. Before the Soviet Union detonated its first atomic bomb on August 29, 1949, these concerns dominated America's Soviet policy.[2]

The Soviet detonation affected American defense policy in several ways. It expedited research on thermonuclear weapons, approved by Truman on January 31, 1950, and it stimulated the formation of an ad hoc study group that outlined the technological parameters for an effective U.S. air defense system against Soviet bomber attack.[3] Moreover, it led directly to the establishment of a special State and Defense departments study group chaired by Paul Nitze, which produced NSC-68, a major reassessment of American national security policy. This paper called for "a rapid and sustained build-up of the political, economic, and military strength of the free world," justified in part by the threat of a surprise Soviet nuclear attack on the United States.[4]

1. "Moscow Embassy Telegram #511: 'The Long Telegram,' " February 22, 1946, reprinted in Thomas H. Etzold and John Lewis Gaddis, eds., *Containment: Documents on American Foreign Policy and Strategy, 1945–1950* (Columbia University Press, 1978), p. 61.

2. This judgment is confirmed by the analysis presented in NSC-20/4, a paper approved by President Truman on November 24, 1948, that served as the official articulation of U.S. policy until April 1950. See "U.S. Objectives with Respect to the USSR to Counter Soviet Threats to U.S. Security," November 23, 1948, reprinted in ibid., pp. 203–11.

3. These developments are authoritatively reviewed in Herbert F. York, *The Advisors: Oppenheimer, Teller, and the Superbomb* (San Francisco: W. H. Freeman, 1976), pp. 41–74, 114–15.

4. See "NSC-68; A Report to the National Security Council," *Naval War College Review*, vol. 27 (May–June 1975), pp. 51–108. The quotation is from p. 108. The report's contribution to American strategic doctrine was discussed in chapter 5.

The acquisition of nuclear weapons by the Soviets confirmed rather than redefined American policy toward the USSR. But this development did raise issues of strategic vulnerability, stimulating support for enhanced U.S. nuclear and conventional forces, the deployment of active defenses, and the building of passive defenses—that is, civil defense.

During the thirty-five years since NSC-68 was written, American defense posture has undergone many shifts, and several of them have been related, though in a complex way, to the quantitative growth and qualitative improvements in Soviet nuclear force deployments. In the variety of force-posture decisions that have been implemented over the years, several interrelated questions have been at issue. What U.S. force deployments are needed to deter a Soviet nuclear attack? How can secure retaliatory forces best be maintained? What mix of conventional and nuclear forces is needed to extend deterrence to America's allies? What role could active and passive defenses play in the protection of civilian populations, urban and industrial targets, and nuclear forces themselves and their command, control, and communication systems?

As technologies have advanced and the Soviet military threat has grown, answers to these questions have changed, and negotiated arms control has been seen as a possible way to stabilize the nuclear competition. But these responses have essentially been in reaction to or justified by the Soviet nuclear program.

Great Britain

The first British atomic weapon test, code-named Hurricane, was carried out at Monte Bello, Australia, on October 3, 1952. It represented the culmination of seven difficult years in British-American nuclear relations in which American policy moved after World War II from an initial position of information denial to one of reluctant and then active cooperation with British nuclear weapon development.

The British government decided in secret to acquire atomic weapons in January 1947, a decision made public in the House of Commons in May 1948. Britain persistently tried to obtain more American support throughout the late 1940s. The British clearly wanted to acquire nuclear weapons to maintain their international status at a time when their vast colonial empire was rapidly disappearing and their economy was in crisis. Britain also believed that nuclear weapons would provide a deterrent against a Soviet attack independent of American control. The weapon program was further seen as a way to maintain a competitive edge on French nuclear weapon developments.

Influential Americans, however, were not keen on assisting the British. Some hoped to establish international control of atomic energy, even after the Baruch Plan failed, and did not want the United States to contribute to nuclear proliferation, even if the proliferator was America's closest ally. Others did not want Britain to retain its Great Power status, seeing a nuclear-armed Britain as a troublesome competitor rather than as a close friend of the United States.[5]

The collapse of any hope of international control of atomic energy and the intensification of hostility in Soviet-American relations swept away much of the American reluctance to cooperate with the British on nuclear matters. After the first Soviet atomic test in 1949 and the outbreak of the Korean War in June 1950, the American mood changed enough to warrant amendment of the McMahon Act in October 1950, permitting substantially increased American support for the British weapon program. Although difficulties remained, and in fact the United States refused a British request to use American atomic-test facilities, the principal barriers to British-American nuclear cooperation were overcome.

Despite a few rough moments since the early 1950s, particularly the Skybolt affair in 1962 and the flirtation with the multilateral force that ended in 1965, relations between the United States and Great Britain concerning nuclear weapons have been remarkably close.[6] Most significantly, American-built submarine-launched ballistic missiles carried on British-built submarines have been the backbone of the British nuclear deterrent force for many years. In addition, the nuclear-force modernization choices Britain faces in the 1980s are intimately linked to U.S. SLBM and cruise-missile programs.

Britain's acquisition of nuclear weapons, therefore, did not affect in any meaningful way the close political relations between the United States and Britain. The impact on American programs has been confined

5. See Margaret Gowing, *Independence and Deterrence: Britain and Atomic Energy, 1945–1952,* vol. 1: *Policy Making* (St. Martin's Press, 1974), p. 265. This work is the official history of the British atomic energy project commissioned by the U.K. Atomic Energy Authority.

6. The Skybolt affair involved the unilateral U.S. cancellation of an American missile program that had provided the only justification for the deployment of a new British bomber built to carry the missile. The multilateral force involved the consideration of multinational crews operating ballistic missile–carrying submarines.

to the sharing of technology and know-how and coordination on selected nuclear policy issues.

France

Relations between the United States and France had had many ups and downs before the French acquired nuclear weapons in 1960. The frictions between Charles de Gaulle and General Eisenhower at critical stages during World War II and the American unwillingness to help French troops during the debacle at Dienbienphu in Indochina in 1954 are two examples of the bitter disagreements that have marked the generally cordial Franco-American political relationship. The turning point in the postwar French relationship with the United States was undoubtedly the Suez crisis of October 1956, in which French, British, and Israeli troops were forced by joint pressure from the United States and the Soviet Union to withdraw from control of the Suez Canal. This experience convinced French leaders once and for all that the United States could not be relied on to serve French foreign policy interests automatically. The plain fact was that throughout the postwar period the United States tried to dominate the policy choices of its allies. But French will, French culture and history, the sense the French have of the role France should play in the world, French mistrust of American policies, and French fear of a militarily strong Germany that could threaten France's security all strengthened resistance to American domination. A sophisticated nuclear-energy establishment and, most important, the leadership exerted by General de Gaulle on his return to power in 1958 resulted in the French acquisition of nuclear weapons.

The French withdrawal from the NATO command structure, the uncertainty of France's commitment to the alliance in the event of war, and the French declaratory policy of *tout azimuth* (being able to attack targets in all directions, presumably including the United States) initially caused significant political upheaval in Franco-American relations and led to a redeployment of NATO's conventional forces and to the assignment of several U.S. ballistic missile submarines for NATO use. But, in the longer view, these were relatively modest disturbances.

The oscillations between cooperation and competition in the relationship between the two powers has continued since France acquired nuclear weapons. De Gaulle was a major supporter of President Kennedy's actions during the Cuban missile crisis and, more recently, France vigorously endorsed the early phases of American détente policy with the Soviet Union and the strategic arms control agreements signed in May 1972. Moreover, French cooperation in NATO military affairs continues in several important respects, although it is carried out unobtrusively. On the other hand, the French were consistently critical of American policy in Vietnam and have pursued foreign policies strikingly different from those of the United States in relation to the Arab-Israeli conflict and sub-Saharan African affairs. Although the French nuclear acquisition has had profound effects on French strategy and force posture, overall it has had strikingly little effect on American defense planning.

The People's Republic of China

After World War II, U.S. policymakers saw communist China as the most powerful threat other than the Soviet Union to American security interests, especially in Asia but also in other regions of the developing world. In the 1950s the Sino-American armed conflict in Korea and several major crises over Taiwan and the neighboring offshore islands occurred. The 1960s began with deepening American involvement in Vietnam, which the United States justified partly as an effort to contain Chinese expansion and to disprove the efficacy of Maoist notions of people's war. The Sino-American relationship, before China acquired nuclear weapons, was wholly adversarial, and those Americans who thought it should be otherwise either had no influence or were summarily discredited if they voiced their views too loudly.

The growing tension between China and the Soviet Union, which opened the way for improved Sino-American relations, was only partly related to China's nuclear weapon policy. Difficulties between the two giant communist states were multidimensional, including the personal animosity between Mao Zedong and Khrushchev; the two countries' differing ideas about how a communist state should be run; incompatible interpretations of Marxist-Leninist ideology and its meaning for developing societies; competition for leadership of the world communist movement; long-standing Sino-Soviet border differences that predate the Russian Revolution; cultural antagonisms between the Russians and the Chinese; and the tensions produced by the geopolitical reality of two powerful states with large populations sharing a 4,500-mile border.

In addition to these difficulties, Soviet leaders were unwilling to make available state-of-the-art expertise and equipment to the Chinese nuclear weapon program, which began in 1954 despite Mao's earlier disparaging remarks about the bomb. Soviet scientists, engineers, technicians, and military advisers all participated in Chinese nuclear weapon research and development until 1960, but the Chinese never received the full benefit of Soviet knowledge, a fact Chinese leaders resented deeply.

It took many years for American policymakers to appreciate the shift in Sino-Soviet relations and its significance. China's acquisition of nuclear weapons stimulated, and was used as an argument for, the deployment of a U.S. ABM system. The U.S. intelligence community anticipated the rapid development of a Chinese IRBM capability, which was used to justify the Sentinel ABM system in 1967.[7] The ABM Treaty, of course, precluded the deployment of this system.

Consequently, there are no U.S. strategic nuclear weapons that were developed or deployed because of any threat, real or imagined, posed by Chinese nuclear forces.

It could be argued that China's acquisition of nuclear weapons was both a cause and a consequence of the Sino-Soviet rift, that this acquisition contributed significantly to China's military independence from the Soviet Union, and that this independence in turn facilitated closer Sino-American relations. But nuclear weapons in this instance seem to have been only secondary to the fundamental axiom of international politics, that the enemy of an enemy is a friend.

Because China could once again become a formidable U.S. adversary, however, it makes sense to understand the evolution of its nuclear forces. Since the explosion of its first nuclear device in October 1964, the People's Republic of China has deployed a small nuclear force suitable for regional use and presumably powerful enough to deter Soviet attack. The modest size of this force and the slow pace of its modernization over the last several years, particularly in contrast to Soviet forces, has meant that Chinese strategic capability has not influenced U.S. deployments and doctrine as much as the United States feared it would in the late 1960s.

Defense analysts do not know why China has not modernized and expanded its strategic forces more rapidly. It is possible that Chinese leaders thought the existing nuclear force was a sufficiently effective

7. The role of these projections in the U.S. ABM debate is discussed at length in Morton H. Halperin, *Bureaucratic Politics and Foreign Policy* (Brookings Institution, 1974), especially pp. 297–310.

deterrent against attack—particularly by the Soviet Union—and that rapid modernization, at least in the short term, was therefore unnecessary. It is also possible that the nation's scarce resources were needed more for domestic than for defense needs; that a shortage of engineers, scientists, and other manpower skilled in high-technology industries dictated a moderate pace; or that ICBMs, modern intercontinental bombers, and sea-based nuclear forces presented technological problems too difficult for the Chinese defense establishment to master. Another reason may be that after Defense Minister Lin Biao's aborted coup d'état in 1971, the political leadership viewed the military with heightened suspicion and tried to weaken it by reducing both the level of defense expenditures and the amount of funds devoted to the strategic forces. Finally, China may have lagged in developing more modern forces to avoid appearing explicitly threatening to the Soviet Union, thereby discouraging a Soviet preemptive strike.

Whatever the causes of China's slow military development, the death of Chairman Mao Zedong in September 1976 brought on events that indicate expansion of military forces in the 1980s and beyond. Mao's successors appear to have consolidated their power with the support of the People's Liberation Army, and the Chinese military establishment has clearly gained influence during this process. The removal of the "radical" element from positions of leadership is one of several indicators that more emphasis will now be placed on the pragmatic goals of economic development and military preparedness rather than on the maintenance of ideological purity through periodic cultural revolutions and class struggles incited by the ruling elite.

Assuming that pragmatists remain in power during the 1980s and that no fundamental rapprochement occurs between China and the Soviet Union, qualitative and quantitative improvements in China's military forces and a greatly strengthened military and technological infrastructure will probably be a high priority. An important question is how willing Chinese leaders will be to compromise China's self-reliance and seek help to achieve these objectives more rapidly than China could achieve them alone.

There are indications that some compromises are already under way. As early as 1975, China completed an agreement with Britain's Rolls Royce, Ltd., to buy supersonic military aircraft engines and production technology that advanced China's propulsion-design technology by several years. Much discussion and some agreements have followed with other Western countries, although major American military assistance will be limited by the U.S. desire to prevent China from becoming too formidable a military power. More recently, China's announced willingness to tolerate private economic initiatives and its professed desire to acquire U.S. nuclear energy technology, possibly even at the price of becoming a party to the Nuclear Nonproliferation Treaty, are indicative of the leadership's commitment to change. The political climate in contemporary China, China's highly adversarial relationship with Vietnam, its deep rivalry with the Soviet Union, and even its uncertainties about the United States (especially in relation to Taiwan) all point toward more rapid modernization of China's military forces in the 1980s and 1990s.

It is unlikely, though, that Chinese weapon modernization will seriously affect U.S. force structure in the near future. The American strategic posture continues to be shaped by the challenges posed by the Soviet nuclear arsenal, which is highly sophisticated and is being modernized rapidly. Even if arms control agreements at some point seriously restrict the quantity and quality of these forces, the United States will no doubt continue to deploy enough strategic weapons to absorb or deter a variety of initial Soviet nuclear attacks. This capability would be more than adequate to deter or, if necessary, respond to a Chinese attack. Unless and until China deploys a modern force capable of delivering nuclear warheads with high accuracy at intercontinental ranges, thereby generating a military threat to the continental United States (as well as to all sectors of the Soviet Union), Chinese military developments will have little direct impact on U.S. force structure.

India

Since India detonated a nuclear device in May 1974, nuclear proliferation in southern Asia has been a growing American concern. However, it is extremely difficult to trace any direct impact this event had on the main character of Indian-American relations or on U.S. defense programs.

India's relations with the United States have been cool at least since the mid-1950s, when Prime Minister Jawaharlal Nehru tried to steer the newly independent India on a course between East and West and when Secretary of State John Foster Dulles enlisted Pakistan, India's archrival, as a member of the anticommunist alliance in southwestern Asia.

Moreover, India's persistent rivalry with China, which erupted in war in 1962, drew India inexorably closer to the Soviet Union, particularly after the Sino-Soviet rift had become serious. An Indian foreign policy emphasizing discord with China and Pakistan and harmony with the Soviet Union—India signed a twenty-five-year Treaty of Peace, Friendship and Cooperation with the Soviet Union in 1971—guarantees friction in Indian-American relations.

This friction intensified in 1971 when the United States favored Pakistan in the India-Pakistan war. The aircraft carrier *Enterprise* and other ships of the Seventh Fleet were sent into the Bay of Bengal, largely to deter India from destroying the sovereignty of West Pakistan after Indian forces had helped transform East Pakistan into the sovereign state of Bangladesh. This U.S. interference, no matter how noble Americans thought it was, left an indelible mark on Indian attitudes toward the United States.

India has had a sophisticated nuclear-energy establishment since the early 1940s and has also felt that American nonproliferation policies were absurdly hypocritical in light of the enormously advanced U.S. nuclear weapon arsenal. These two factors made India's decision to detonate an atomic bomb in 1974 relatively easy.[8] Since the detonation, many controversial issues have emerged in Indian-American relations, particularly with respect to American nuclear fuel supplies for the Indian reactor at Tarapur. The relationship between the two powers has remained cool, with occasional bouts of acrimony, but is far from adversarial. The Indian nuclear explosion and the prospect this raises of a full-fledged Indian nuclear force has yet to influence seriously cultural, economic, scientific, or political ties between India and the United States.

Israel

The United States has been Israel's principal supplier of arms and economic assistance only since the 1967 Arab-Israeli war, when the Middle East policies of several European countries, particularly France, shifted. Since then, American-Israeli relations have been complex. The United States has tried to influence Israeli behavior to promote American

8. Indian perspectives and capabilities are described in detail in Onkar Marwah, "India's Nuclear and Space Programs: Intent and Policy," *International Security*, vol. 2 (Fall 1977), pp. 96–121.

strategic interests in the region—that is, to contain Soviet influence, improve relations with Arab oil-producing states, and preserve Israeli security—while Israel has tried to manipulate American military, political, and economic support and diplomatic involvement in the Arab-Israeli conflict to promote Israel's perceived security interests.[9]

What is striking about the relationship, however, is the extremely limited and subtle influences Israel's prospective acquisition of nuclear weapons has had on American policy. Rumors of Israeli nuclear weapon acquisition have been widespread since the late 1960s and are usually associated with weapon-grade material that might be produced at the Dimona nuclear research facility in the Negev desert.[10] But while American leaders have been aware of the Israeli nuclear program, there is no evidence that this knowledge directly affected U.S. policy until the 1973 Arab-Israeli war. William B. Quandt, for example, in analyzing the 1969–70 Middle East initiatives advanced by Secretary of State William P. Rogers, observes that "no one knew quite what to do about the Israeli nuclear option, but it added to the sense that the Middle East was too dangerous to ignore."[11] And Nadav Safran, in analyzing the unfolding developments of the October 1973 war, asserted that "Kissinger, along with a few people at the top government echelons, had long known that Israel possessed a very short nuclear option which it held as a weapon of last resort, but he had not dwelt much on the issue because of the remoteness of the contingency that would make it relevant."[12]

Perhaps the only indirect influence of Israel's nuclear option on American policy before 1973 was the impetus it created to provide Israel with sophisticated conventional weapons, a way to lessen the Israeli incentive to proceed with a nuclear program.[13] During the 1973 war, however, American willingness to resupply Israel with vast amounts of arms early in the conflict was tied directly to the fear that Israel, seeing

9. Two excellent analyses of this complicated relationship with quite different perspectives are Nadav Safran, *Israel: The Embattled Ally* (Belknap Press of Harvard University Press, 1978); and William B. Quandt, *Decade of Decisions: American Policy toward the Arab-Israeli Conflict, 1967–76* (University of California Press, 1977).

10. One rather explicit but unconfirmed report of the origins and development of the Israeli capability is contained in "How Israel Got the Bomb," *Time*, April 12, 1976, pp. 39–40.

11. Quandt, *Decade of Decisions*, p. 80.

12. Safran, *Israel*, p. 483.

13. This is an example of the "dove's dilemma," the urge to provide a threshold nuclear state with advanced conventional weapons to dissuade it from becoming a nuclear weapon state.

itself on the verge of defeat, would resort to the use of nuclear weapons to defend itself.

The ambiguous status of the Israeli nuclear program has not affected the strengths or the difficulties of Israel's relationship with the United States, nor has it affected U.S. defense programs. It has, however, influenced American diplomacy in a time of crisis and presumably would do so again in the future. The pressures for the United States to come to Israel's defense and to try to help resolve its conflicts in the event of war are especially intense, because the price of American procrastination or failure could be nuclear war in the Middle East.

Conclusions

In general, then, the acquisition by various countries of nuclear weapons has not fundamentally changed their political relationship with the United States. Except for the Soviet Union, no new nuclear state has significantly affected U.S. defense programs or policies, and American interest in bilateral nuclear arms control has been confined to negotiations with the Soviet Union. Finally, a conventional conflict involving a nonnuclear ally (Israel) prompted the United States to intervene in ways it otherwise might not have to forestall the possible use of nuclear weapons.

The Future

If the future were to be like the past, there would be remarkably little to worry about, even as nuclear weapons proliferate to other countries. American foreign policy would have to be adjusted only slightly, and American defense programs would remain tailored solely to the Soviet threat. The likelihood of any country using nuclear weapons would remain low, and American diplomacy would be called upon principally to prevent near-nuclear states from using nuclear weapons as a last resort. Two emerging conditions suggest, however, that the future will not resemble the past. The first is the intensification of the Soviet-American competition; the second is the changing character of regional security.

The Significance of Intensified Soviet-American Competition

Soviet-American relations have warmed and then cooled several times since the end of World War II—from the immediate glow of victory over Germany to the confrontation over the Berlin blockade; from flirtations with "Open Skies agreements" and the "spirit of Camp David" during the Eisenhower administration to the Cuban missile crisis a few years later; from the birth of détente in the early 1970s to its collapse by the end of the decade.

The present phase of the Soviet-American relationship, marked by competitiveness and acrimony, will most likely continue for some time; its implications for U.S. defense policy and nuclear proliferation are substantial. It has already been noted that political pressures as well as technological advances are leading to the development and deployment of high-accuracy systems that can carry out selective strikes with limited secondary damage. Cruise missiles are a notable example. These systems, armed with conventional and nuclear warheads, will surely be a significant part of the Soviet and American military arsenals by the late 1980s. As the Soviet-American military competition intensifies, the United States may relax its restraint on the transfer of cruise missile technologies to its allies. In time, several countries may obtain these technologies (including high-specific-impulse engines and sophisticated, miniaturized guidance systems). Cruise missiles could then become attractive nuclear-delivery vehicles for new nuclear states, especially in view of the missiles' low cost compared with that of aircraft or ballistic missiles.

It is not just a matter of U.S. and Soviet nuclear weapon programs perhaps stimulating nuclear proliferation. The problem is more complex. Some of the Soviet and American systems that are being considered for deployment might be justified on the grounds that they could strike nuclear forces of hostile new nuclear states or could defend the superpower homeland against attacks by such states. The prospect or proof of nuclear proliferation could even be useful in the American debate in favor of systems directed primarily at the Soviet threat. Moreover, the deployment of a wide variety of such systems by the United States and the Soviet Union would give credibility to the arguments of proponents of nuclear weapon acquisition in nonnuclear states. These proponents would have additional evidence supporting their contention that the superpowers are indeed hypocritical about nuclear arms control and in fact pose a menace not only to each other but also to nonnuclear states.

Lack of progress in bilateral arms control negotiations and the possible unraveling of existing agreements, especially the ABM Treaty, would no doubt cripple the prospects for multilateral nonproliferation initiatives.[14] The Nuclear Nonproliferation Treaty would be weakened, and there would be little prospect of the Treaty of Tlatelolco (the Latin American Nuclear Free Zone Treaty) entering into force for Brazil or Argentina or of other nuclear-free-zone arrangements being completed. Lack of progress on these diplomatic fronts would, in turn, reduce the political barriers to acquiring nuclear weapons in many threshold countries.

It has often been observed that the Soviets have the most to lose in a world of many nuclear powers because the vast majority of these countries would be hostile to them. Therefore, the Soviet Union has an incentive to cooperate with the United States on nuclear nonproliferation policy, regardless of the general nature of the relationship. Indeed, it is often overlooked that nuclear nonproliferation is one of the most fruitful areas for Soviet-American cooperation.[15] However, this perspective is probably too apolitical to be useful, and it may not be feasible for the Soviet Union to cooperate on any front if the bilateral relationship becomes wholly confrontational.

The Changing Nature of Regional Security

The pace of nuclear proliferation might not be as leisurely in the future as it has been in the past. If proliferation chains develop,[16] they will probably affect American foreign and defense policies. Such chains

14. Most proponents of American abrogation of the ABM Treaty are dominated by concerns about the Soviet threat. But the demonstration of nuclear weapon proliferation by one or more states hostile to the United States would substantially strengthen the argument for abrogation in U.S. domestic political circles.

15. The United States and the Soviet Union were the key architects of the Nuclear Nonproliferation Treaty, the Soviets being particularly motivated to close off West German nuclear weapon options. Moreover, in the fall of 1984, at a time of deep suspicion in U.S.-Soviet relations, quiet diplomacy was being conducted by the two governments to coordinate their respective nuclear nonproliferation policies.

16. A proliferation chain is a rapid sequence of countries' acquiring nuclear weapons, each acquisition a result of the preceding one.

would most likely produce more pairs of contiguous nuclear rivals—Israel and Syria, Egypt and Libya, India and Pakistan, North and South Korea, Argentina and Brazil, the People's Republic of China and Taiwan—increasing both the incentives for preemption (illustrated by the Israeli attack on Iraq's Osirak nuclear reactor in June 1981) and the likelihood of nuclear weapon use in the escalation of a conventional war. Contiguity has always been a useful indicator of the likelihood of conflict in international relations; there is no reason to suspect that it will not be an important determinant of nuclear conflict as well.[17]

Consider the following prospective list of new nuclear states categorized according to their political relationship with the United States:

1. *Major ally:* Federal Republic of Germany; Japan
2. *Potential or actual regional ally:* Pakistan, South Korea, Taiwan
3. *Neither ally nor adversary:* Argentina, Brazil, South Africa, Yugoslavia
4. *Adversary of regional ally:* Iran, Iraq, Libya, Syria

If West Germany or Japan decides to acquire nuclear weapons because of dramatic changes in either its domestic affairs or in international conditions, it is vitally important that the United States make every effort to retain strong political relations with both countries. The most decisive setback to America's international position, aside from or perhaps even including a Sino-Soviet rapprochement, would be the loss of a nuclear-armed Germany or Japan as a military, economic, and political ally. However, if either of these countries acquires nuclear weapons, there is no inherent reason why acquisition would have to affect adversely American policy any more than the British and French acquisitions have.

U.S. failure to accommodate to such a transition would have profound consequences not only for international political alignments but for all facets of American defense. A nuclear-armed Germany or Japan in an adversarial relationship with the United States would surely stimulate a major growth in U.S. offensive nuclear forces, ballistic missile defenses to protect nuclear retaliatory forces and possibly even population

17. It could be argued, of course, that contiguity would be a deterrent to nuclear weapon use if it was appreciated that destruction of thy neighbor led through radioactive fallout to destruction of thyself. More generally, for an argument that the gradual spread of nuclear weapons would promote deterrent balances within regions, see Kenneth N. Waltz, *The Spread of Nuclear Weapons: More May Be Better,* Adelphi Paper 171 (London: International Institute for Strategic Studies, 1981).

centers, and air defenses and civil defense programs. A new nuclear enemy would also affect dramatically the deployment patterns of U.S. general purpose forces.

The acquisition of nuclear weapons by Pakistan, Taiwan, or South Korea would probably affect American foreign policy more than U.S. defense programs. An Indian-Pakistani nuclear standoff would challenge American diplomacy to reduce tensions in the region, limit Soviet influence in South Asia, and, if possible, help forestall the use of nuclear weapons should conventional war break out between the parties. A South Korean nuclear force could rupture South Korea's security arrangement with the United States, since U.S. officials could claim that the weapons' deterrent value would preclude the need for American troops in South Korea. Possession by Taiwan of nuclear weapons would raise serious problems for the United States. Although few Americans would want to risk nuclear war with China over Taiwan, there is a strong desire in the United States to maintain Taiwan's territorial integrity.

If countries in the third category—Argentina, Brazil, South Africa, and Yugoslavia—acquired nuclear weapons or emulated the Israeli model of fostered ambiguity, effects on U.S. security policy would probably be minimal. Bilateral relations would probably worsen in the short term: the United States might want to punish a new nuclear state by imposing economic or political sanctions. But American leverage in each of these countries is quite limited. Once the United States realized that its regional interests, particularly trade and economic investment and the containment of Soviet influence, were not directly jeopardized by nuclear proliferation, the United States would probably adjust with little difficulty.

Nuclear proliferation in countries in the fourth category—Iran, Iraq, Libya, and Syria—would pose fundamental problems for American policymakers. The likelihood of nuclear weapon use would be perceived to be great, Israel's and possibly Egypt's survival would be in jeopardy, the Arab-Israeli armed conflict would probably reignite, heightening the vulnerability of oil fields and tankers in the Persian Gulf, and nuclear terrorism could become reality. The United States would have incentives to become directly involved in the region in ways it never has before—by stationing permanent troops, formulating explicit security guarantees, and even contemplating preventive strikes against nuclear storage depots or delivery vehicles. Such actions would raise the likelihood of a Soviet-

American confrontation in the region. There seems little doubt, therefore, that a group of nuclear-armed rivals in the Middle East would have a far greater effect on U.S. security policy than Israel's behavior alone has had.

The Basic Options

The United States will be forced to choose among a number of basic options in adapting to a world of nuclear powers. One option is that of malign neglect. America could permit nuclear proliferation to proceed without attempting to use its considerable political and economic influence to slow the process. It could, to a very great extent, withdraw from the international arena, passively witness the dissolution of its structure of alliances, and remain deliberately aloof from nuclear and nonnuclear conflicts among other nations. It could turn inward, concentrating instead on the maintenance of its massive military arsenal for deterrence purposes while building an elaborate array of air and ballistic missile defenses to protect itself as well as possible from a nuclear attack on its continental homeland.

There is virtually no prospect of this option being adopted, however. The sense of the United States as a superpower continues to be the dominant perception, not only of the vast majority of Americans, but of most of the peoples and governments around the globe as well. With this perception of America's status, there is, at least implicitly, a set of concomitant rights and responsibilities that the United States is expected to exercise. As the most powerful military nation of the Western democracies, it is looked upon as the principal counterweight to the expansion of Soviet influence. The strength of the American domestic economy, its extraordinary technological sophistication, and its pervasive international reach necessarily involve the United States in a web of interdependent political and economic relationships with scores of nations whose governments, their rhetoric notwithstanding, seek such relationships. And even states that have often viewed themselves as potential adversaries of the United States—China, some of the socialist states of Eastern Europe, a number of Arab nations—have, in their own national interests, argued against American retrenchment and sought American diplomatic assistance. Consequently, there are no significant voices in the United States or abroad that are likely to persuade America to adopt the option of malign neglect.

A second option for the United States would be to promote nuclear realignment. This would involve a conscious and overt American effort to scrap its existing security arrangements in favor of creating a ruling elite composed of the nuclear weapon states. Since it is already the case that every permanent member of the United Nations Security Council is a nuclear "have" nation, there is precedent for such an approach. The aim of this realignment would be to formalize what is an implicit fact in world affairs: nuclear weapons are significant and they automatically elevate the status of those states that possess them. By accentuating the differences between the nuclear haves and the nuclear have-nots, a commonality of interests would be fostered among the nuclear nations, thereby reducing the likelihood of interstate conflict among the nuclear powers. Indeed, since its nuclear explosion, India has at times conveyed the impression that it is due certain privileges now that it has entered the nuclear club. By granting such privileges and underscoring the nuclear-nonnuclear distinction, the United States could promote the idea that the nuclear badge had become the lone symbol of the powerful in international politics. Over time this condition could, ironically, impose order on rather than create chaos within the international system.[18]

This option, however, has little to recommend it. As a consequence of the deep political divisions among the nuclear powers, a compact among them would be extremely fragile at best, and they would be almost certainly incapable of adopting a coordinated policy to halt the spread of nuclear weapons. On the contrary, promoting the status of nuclear weapons in this way would surely encourage nuclear proliferation rather than deter it, placing in many hands the ability to initiate nuclear war. The result would be large numbers of vulnerable forces with inadequate command and control systems, subject to seizure by

18. This notion of nuclear realignment providing international stability is a variation of the concept of a unit veto system in which it was hypothesized that the possession of nuclear weapons by all nations would promote a condition in which each state could successfully deter an attack upon its homeland initiated by any other state. See Morton Kaplan, *System and Process in International Politics* (Wiley, 1957), pp. 50–52, and his more recent thoughts on the subject, "The Unit Veto System Reconsidered," in Richard Rosecrance, ed., *The Future of the International Strategic System* (San Francisco: Chandler Publishing Co., 1972), pp. 49–55.

national and subnational groups. Preeemptive strikes would be more likely. An increase in regional conflicts initiated by nuclear powers against nonnuclear neighbors could be anticipated. And because the superpower rivalry would not be muted, there would be a substantial probability that in regional conflict situations the Soviet Union and the United States would support opposing parties, thereby running the risk of frequent nuclear confrontations. American security interests would therefore not be at all well served by a realignment of nations that underscored the primacy of nuclear weapons.

A third option is to adopt a policy of confrontation politics. In many ways, this would be the opposite of the second option. Rather than promoting the elite status of the nuclear states, and thereby encouraging proliferation and instability, the United States, acting either alone or in concert with the Soviet Union, could behave as a nuclear bully, imposing or threatening to impose severe sanctions against any state that either appeared intent on acquiring nuclear weapons or had in fact done so. The range of sanctions could be quite extensive: the severing of diplomatic relations, the cancellation of trade and assistance agreements, the provision of nuclear weapons to the regional rivals of new nuclear states, and "surgical strikes" against the new nuclear facilities and associated delivery vehicles. This option would amount to the overt use of United States political, economic, and military power to prevent near-nuclear states from crossing the nuclear threshold.

The likelihood that this approach will be adopted, except in rare instances, is remote. The will of the executive branch to use confrontation politics and perhaps military force in order to stem nuclear proliferation in regions distant from those areas widely regarded as bearing quite directly on U.S. national security interests (that is, Western Europe, Israel, Japan, and South Korea) is not very great. The Vietnam experience, applicable or not, would most likely be invoked, and the support for such actions in Congress and among the American electorate is unlikely to be forthcoming: opposition and denunciation is the far more probable reaction. Moreover, threats of coercion, if proved not to be credible, would be counterproductive in that they would reveal the weakness of U.S. nonproliferation policy. There is, in addition, the problem of applying the policy in a consistent and evenhanded way. Would the United States be willing to implement the same sanctions against both a nuclear Israel that felt threatened with annihilation and a nuclear Iran that had acquired these weapons as part of an overall policy to seek hegemonic control over the Persian Gulf states? Hardly. To determine a set of sanctions that would be appropriate in all cases to the perceived threat while commanding both domestic and international support would be enormously difficult. As a consequence, this option is unlikely to be the principal path taken by the United States to cope with nuclear proliferation.[19]

A fourth option might be termed equality promotion. Many students of the nuclear proliferation problem believe that its underlying cause rests in the fundamental inequality among states–inequality in economic, political, and military terms particularly between the rich states of the North and the poor states of the South. This much-discussed gap between North and South must be narrowed, it is argued, if the incentives to acquire nuclear weapons are to diminish. To tackle the problem from this vantage point would place great demands on the United States. As the world's foremost economic power, the United States would be called upon to lead the way in redressing the imbalance between North and South. This might mean a return to the massive American foreign aid programs of the 1950s and 1960s, transfers of whole industrial plants, infusion of high-technology products including advanced digital computers, the establishment of management training programs and other educational curricula from preschool through graduate studies—all this to scores of nations in an effort to stimulate economic and social development in much of the third world.

These efforts would need to be supplemented, in many instances, by military assistance. A large number of nations would seek to obtain modern, sophisticated conventional arms of the quantity and quality that, in recent years, the United States has provided to Egypt, Saudi Arabia, and Israel, and they would not have the funds to pay for them. At the same time, the United States would have to reduce significantly its nuclear arsenal, either unilaterally or through joint negotiations with the Soviet Union. Reductions in the number of delivery vehicles and

19. This is not to imply that sanctions would have no place in American nonproliferation policy. The ability of the United States to pressure South Korea in 1976 to terminate its contract with France for a chemical reprocessing plant indicates that confrontation politics can be applied and can be successful. This is particularly true in situations in which the United States has substantial economic and military investments that it could credibly threaten to withdraw. But the thrust of the argument is that such situations are not numerous and that, consequently, confrontation politics cannot dominate U.S. policy.

nuclear warheads and a substantial slowing of the pace of weapon modernization would all have to be clearly demonstrated before many of the nuclear threshold nations would feel that North-South inequality had been narrowed sufficiently to warrant their forgoing the nuclear option. Failure to take these steps would only confirm the view currently held in many third world states that the acquisition of nuclear weapons is the only option available to them to equalize the power imbalance between the rich and the poor.

This path, like the others, is unlikely to be taken, and for a number of reasons. Inequality has been an inherent characteristic of both human affairs and international affairs throughout the ages. Except in the rarest of circumstances, the rich have never willingly taken great strides to equalize material imbalances. Rather, the poor have remained poor, or they have grown rich through their own efforts, or through good fortune, or because they used force to obtain wealth from the rich. It is extremely unlikely that the rich nations generally or the United States in particular will deviate from this pattern. Moreover, it is not at all clear that, if such inequality gaps were bridged, the effect on nuclear nonproliferation would be favorable. For it is becoming increasingly apparent that the term *North-South conflict* is exceedingly imprecise as an explanation of the forces stimulating nuclear proliferation. The South, in fact, is composed of a highly heterogeneous group of nations that differ markedly in all measures of national wealth. And for most nations of the South, it is a neighboring member of the same region that is its principal rival and its principal threat. Surely Argentina seeks a nuclear capability far more because Brazil may soon acquire nuclear weapons than because Britain and France already have them.

To be sure, the United States can be expected to take some role in the years ahead to ameliorate North-South relations. But this will fall far short of achieving a condition of equality among nations. The view is strongly held in U.S. policymaking circles that the incentives for nuclear proliferation rest more with issues of national prestige and regional conflicts than with matters of economic and military inequality. America is, therefore, not about to embark on an enormously costly program of economic and military assistance nor to initiate substantial unilateral reductions in its nuclear arsenal that might provide the Soviet Union with political or perhaps even military advantages.

A fifth option that might be termed adaptive continuity is the likely approach that the United States will follow in coping with a world of nuclear powers. This option involves using a mix of strategies tailored to specific aspects of the nuclear proliferation problem and adapting, as most governments do, pragmatically and incrementally to changing circumstances as they unfold.

Attempts to control the spread of nuclear weapons will continue to move along essentially two lines: political-military strategies and energy-related strategies. These strategies may be summarized as follows.

Political-Military Strategies

Strengthen the Nonproliferation Treaty. This remains the principal international legal instrument for retarding nuclear proliferation. Unless or until its legitimacy collapses in the face of a large number of states violating or withdrawing from it, the United States will do what it can to see that it is retained and that its effectiveness is improved. The treaty will therefore receive overt and high-level support from the United States, which will continue to attempt to induce nonmembers to ratify it.

At this stage the treaty is a necessary but not a sufficient condition for nuclear proliferation to be controlled. It is of great symbolic significance, and its collapse would signal that the effort to control proliferation had failed, thereby further stimulating the nuclear spread. The United States is likely to move in the direction of attempting to create incentives for nonmembers to join the treaty, such as making the transfer of U.S. nuclear technology conditional on the recipient nation being a member and accepting full safeguards as specified in the treaty. It is unlikely, however, that the United States will move so far as to offer amendments to the treaty, for fear that the amendment procedures could open the way for some states to withdraw.

Adopt pledges of nonuse of nuclear weapons against nonnuclear weapon states. There is the prospect, although it is not very great, that the United States will consider some form of pledge concerning the nonuse of nuclear weapons. Great caution must be exercised in this area because such pledges could create serious doubt about the credibility of American security guarantees to its allies in Europe and East Asia. It is possible that a nonuse pledge could be formulated with the provision that it would not apply to nonnuclear states that assist one or more nuclear weapon states in aggressive actions against other nuclear weapon states or their allies. Indeed, it could be accompanied by explicit assurances to the Federal Republic of Germany and Korea concerning American intentions. If such a pledge proved satisfactory to America's allies and were endorsed by the other nuclear powers as well, it might reduce the likelihood of nuclear weapon use, it would be an indication of the diminishing political utility of nuclear weapons in the foreign policies of the major powers, and it would be significantly reassuring to nonnuclear states.

Establish nuclear free zones. There remains enormous difficulty in establishing nuclear free zones, as evidenced by the Treaty of Tlatelolco, and it is precisely in those areas of the most intense regional conflicts—the Middle East, the Indian subcontinent, and Korea—that the inevitable variations in military potential among the protagonists make the chances for reaching such agreements least likely. Moreover, the nuclear free zone approach will become hopelessly utopian as nuclear weapons spread. The best the United States could hope to do would be to assist where feasible in the settlement of regional disputes among nonnuclear powers, thereby diminishing the need for these states to acquire nuclear weapons.

Maintain and enhance security guarantees. It can be anticipated that the principal stimulants to nuclear proliferation will continue to be threats to the security of nonnuclear states. By maintaining and enhancing security guarantees, the United States could both constrain the options of the nonnuclear states and provide assurances to them that would diminish the incentives for these states to develop independent nuclear weapon capabilities. Whether the United States will be able to establish and preserve new guarantees is difficult to assess. At present, long-standing commitments involving the United States are being increasingly (not decreasingly) questioned, and the prospect of extending these arrangements in a credible fashion to other threshold countries appears to be quite slim. Moreover, in cases where the establishment of security guarantees might be possible, the conditions demanded by the nonnuclear states in terms of economic and military aid may prove to be unacceptable to the United States. Nonetheless, as nuclear weapons continue to spread, the maintenance of security guarantees is a most logical course for the United States to pursue.

Impose sanctions against new nuclear states. Because of the lack of support both at home and abroad for U.S. armed intervention, except where its vital interests are concerned, there is little prospect that the use of military sanctions will be a significant U.S. tool in coping with nuclear proliferation. Political and economic pressures and sweeteners are likely to be applied, however, to persuade nonnuclear allies to remain nonnuclear. Should one or two nations cross the threshold, the United States may well try to deter others by heavily penalizing the new nuclear states. But these penalties will be country-specific, limited in scope and duration, and unlikely to include military sanctions.

Energy-Related Strategies

Establish supplier agreements to regulate the transfer of nuclear technology. Nuclear proliferation can be constrained by minimizing the opportunities for national governments to acquire weapons-grade material. This fact will be as true in a world of many nuclear powers as it is today. The United States, by reaching agreement with the other nuclear supplier nations not to sell or assist in the development of fuel reprocessing plants or uranium enrichment facilities, could impede further nuclear proliferation, no matter what its level, at any time. By agreeing to provide and subsidize fuel cycle services at substantial economic advantage to recipient states (compared with those states having their own facilities), the United States could seek to enter into agreements that guarantee the security of supply of energy resources to these recipients. In cases where nuclear facilities are provided, assurances must be obtained that the spent fuel will be returned to the supplier. There are considerable difficulties inherent in this strategy in terms of managing alliance relationships and coping with the instabilities of a nuclear cartel. But the benefits to be derived are worth the effort. This approach should continue to be an effective nonproliferation strategy in the future.

Exercise unilateral restraint in nuclear energy programs. It will remain highly desirable for many years to come that the United States refrain from taking steps with respect to its own nuclear energy programs that would increase the likelihood of the spread of weapons-grade material. Delaying both permission for plutonium recycling and the decision to produce fast-breeder reactors is consistent with this objective. Should the United States move ahead with both these developments, weapons-grade material would spread extremely rapidly and the battle to retard nuclear proliferation would be lost. Those who support plutonium recycling and the production and sale of fast-breeder reactors point to unproven economic arguments to substantiate their case. Unless and until extraordinarily compelling economic arguments can be made in support of these programs, there is every reason to delay them.

Establish multinational nuclear facilities. In the years ahead, the United States may well take the lead in establishing multinational nuclear facilities as alternatives to national facilities. These facilities could perform fuel cycle operations from uranium enrichment to waste disposal and could vary in organizational structure from a bilateral supplier-recipient arrangement serving the fuel cycle needs of only one nation to a system of multiple ownership and management serving an entire region. The establishment of such facilities would increase the economic and political costs of acquiring national nuclear facilities and would increase the effectiveness of safeguards against diversion of materials, since each of the participating nations would serve as a monitor against diversion by others. Although multinational facilities run the risk of stimulating reprocessing activities and serving as agents for the transfer of technological know-how, these risks are worth taking. Serious consideration by the United States of converting one or more of its national facilities into a multinational facility is a likely prospect in the years ahead.

Increase financial support to the IAEA. To prevent the diversion or theft of materials from national facilities, the safeguards program of the International Atomic Energy Agency (IAEA) will undoubtedly require expansion, and increased efforts will need to be made to provide physical protection of these facilities. In effect, such measures can be achieved only with substantial technical and financial assistance from the United States. While care should be taken to ensure that American assistance does not so dominate the agency that it comes to be perceived as a tool of U.S. policy, sufficient aid must be provided to generate an effective IAEA safeguards program. This program is a necessary, although not a sufficient, condition to control nuclear proliferation in the years ahead.

Nonproliferation policies will continue to be implemented in a world of many nuclear powers. In the face of the nth power acquiring nuclear weapons, the United States will act to prevent the $n + 1$ power from taking the same step.

There is, however, one significant caveat with respect to the option of adaptive continuity, and this concerns the pace of the nuclear spread. The United States has, until now, been able to adjust to nuclear proliferation because its pace has been gradual and the spread has been largely restricted to the major powers, which have exercised extreme caution in their handling of the weapons. Should proliferation accelerate, technical deficiencies in the new nuclear forces and the prospect that some national leaders might act irresponsibly will make the world a far more dangerous place and will adversely affect American security interests. A major shock to the system—for example, the sale of a nuclear weapon, the use of a nuclear weapon in anger, or the widespread availability of laser isotope separation technology and expertise permitting off-the-shelf bombs to be acquired by many states within a one-year period—would so suddenly and dramatically alter the nature of international political relationships that it is highly improbable that the United States, under such circumstances, would retain an incremental and cautiously pragmatic approach to the problem.

Should a shock to the system materialize, the United States is likely to move aggressively to halt further nuclear proliferation, using coercive and military measures where necessary. The building of specialized offensive and defensive forces tailored to meet the threat posed by small nuclear-armed states would be encouraged. In some instances, the United States might share its technological expertise with new nuclear states to improve their command, control, and communication systems so as to minimize the likelihood of accidental or unauthorized use of nuclear weapons. While a fundamental rethinking of defense policy and alliance relationships might well be stimulated, it is highly probable that sanctions against new nuclear states would dominate the American reaction.

Whether U.S. security in a world of nuclear powers can be maintained utilizing any of the measures suggested above is an open question. Until the option of adaptive continuity is proven to be ineffective, however, it will constitute the American response.

In general, the greater the spread of nuclear weapons, the more nuclear states will be perceived as potential U.S. adversaries, and the more powerful the incentive will be to develop new weapons and defenses. American policymakers must both compete with the Soviet Union in nuclear armaments and simultaneously pursue nuclear nonproliferation.

The pace of proliferation may be slowed somewhat by its perceived costs to the proliferator, such as heightened regional tensions or endangered security arrangements with the United States. In any case, American policy must continue to rest on strengthening nonproliferation; reducing the incentives for states to acquire sensitive nuclear facilities (for example, uranium enrichment facilities, breeder reactors, and nuclear fuel reprocessing plants) from which they could acquire weapon-grade material; and maintaining close security ties with threshold states when feasible. However, the imperatives driving the United States to compete with the Soviet Union in nuclear weapons will continue to undercut U.S. efforts to stem the tide of nuclear proliferation.

CHAPTER NINE

The Search for Security

The right to search for truth implies also a duty: one must not conceal any part of what one has recognized to be true. Albert Einstein

IN SPITE OF intractable problems, Americans have a responsibility to do what they can to ensure their survival in a world that holds more than 60,000 nuclear weapons. It is their obligation to see the problems clearly and manage them in ways most likely to avert nuclear war.

The history, geography, and social and political structure of the Soviet Union reveal a pervasive sense of insecurity within the Russian psyche. The absence of security in every aspect of Russian life is striking. Insecurity extends from the borders of the state to interpersonal relations, from the uncertainty of political authority to the uncertainty of whether there will be enough to eat. With this insecurity has come an instinct for hierarchy and dominance as the basis for the survival of both the worker and the government. For the Soviet Union nuclear weapons are the supreme guarantor of the security of a state that has never really known security, a symbol of communism's technological achievements, and a means of implementing the expansionist foreign policy goals of the ruling elite.

Therefore, the Soviet government must be treated with firmness but without belligerence. President Kennedy's dictum that "we must never negotiate out of fear, but we must never fear to negotiate" is sound and practical advice. To show weakness to Soviet leaders on the one hand or to try to isolate them from the international community on the other are both dangerous approaches to the dilemma of U.S. relations with the Soviet Union.

But it is difficult for the United States to maintain an evenhanded approach because its citizens have a split personality when it comes to nuclear weapons. Americans cannot make up their minds about these enormously destructive devices. They are technologically sweet, a symbol of American expertise and strength, and they protect Americans and their allies from communist (essentially Soviet) aggression. But

Americans feel guilty about their use of these weapons in the past. They fear that the nuclear competition will get out of hand, if it has not already. They hope that the nuclear arms race can be "solved" or made to disappear. Americans see international politics as a set of problems with solutions rather than a series of ever-changing conditions that can be managed for better or worse but that cannot be made to disappear.

As a result there is strong support in the United States for both a strong military second to none and for arms control agreements with the Soviets. Futhermore, Americans seem impatient with the permanence of the U.S.-Soviet rivalry and tend to swing between extremes of optimism and pessimism and cooperativeness and hostility in a fruitless search for the answer to the Soviet riddle. However, to cope effectively with the dilemma, U.S. policy needs to be consistent and reflect the realization that the United States cannot fundamentally change the Soviet political and economic system.

In terms of military technology, a change of potentially revolutionary implications is approaching: the era of high-accuracy weapons. In previous eras of warfare, most shots missed their targets. For the rest of this century weapon systems will be able to hit what they are intended to hit and not damage much of anything else. In warfare, premium value will be placed on mobility, deception, and defense. Eventually all fixed targets and many mobile targets will become vulnerable to attack. Accuracy will no longer be a function of range, and the distance of the attacker from the target will no longer determine the attacker's ability to find and destroy it.

High-accuracy weapon systems have not yet seriously threatened either superpower's retaliatory forces. Both sides could detonate a few or a few thousand nuclear warheads on a variety of military, industrial, and civilian targets after absorbing an initial attack. Neither side is close to having a disarming first-strike capability. Therefore, as long as rationality prevails and accidental launches are avoided, mutual deterrence will keep the nuclear peace. Indeed, despite the public rhetoric equating the proliferation of U.S. and Soviet systems with impending nuclear doom, such proliferation will probably prevent nuclear doom. In particular, the prospective deployment of large numbers of relatively invulnerable but slow cruise missiles—second-strike missiles—will enhance deterrence, at least marginally.

There are so many diverse nuclear systems that opportunities for nuclear coercion seem nil for the present. Moreover, since offensive systems dominate the scene, the prospects for an adequate defense by either superpower against a premeditated attack by the other, even with small numbers of weapons, is very low. The variety of ways the offense can fool or overcome the defense raises serious questions about the ability of command, control, communication, and intelligence systems to function in a nuclear environment. Currently C^3I systems are extremely vulnerable to attack, and the likelihood that existing systems could function in a nuclear environment is slim. It is unlikely that in this decade or well beyond, technological advances, especially directed-energy weapons or the extension of the military competition into space, could upset the current offense-dominated strategic balance.

Primarily because of technological advances, U.S. nuclear war plans have been consistently revised to reflect the accuracy and flexibility of U.S. strategic forces. Contrary to what most citizens seemed to believe, recent changes endorsing greater flexibility in targeting of the forces were incremental and did not reflect fundamental shifts in policy. Indeed, historically there has been great disparity between the statements and writings of senior officials (declaratory policies) and actual war plans (employment policies) for the use of strategic forces. Since the 1950s U.S. leaders have always placed greater emphasis on military than on civilian targets and on limited rather than spasm responses, although declaratory policies have often stressed the opposite. Both U.S. and presumably Soviet strategic doctrine place far more emphasis on destroying the opposing forces and the war-making potential of the adversary rather than its population centers. It is not clear, though, the degree to which such doctrines would actually guide policymakers in a real nuclear war.

In the area of arms control, none of its objectives as originally conceived—reducing the likelihood of war, the damage should war occur, and the resources devoted to war—are being met by current arms control negotiations and agreements. Instead, a complex set of political and military motivations has moved both sides to pursue arms control policies based on aspirations of unilateral advantage. Moreover, no major U.S. decision about the deployment of offensive strategic forces has been significantly affected by any arms control negotiation or agreement. A more accurate term for arms control, then, is threat control. If each side could restrict a key threat it posed to the other, regardless of the asymmetries involved, political support for arms control could be sustained, and the ensuing agreements, whether explicit or implicit, would make a meaningful contribution to strategic stability.

With respect to its European and Japanese allies, the United States is in a difficult period. Differences of opinion about the extent of the Soviet threat as well as strains from economic competition and changing generations' attitudes are undermining alliance cohesion. New precision technologies will make it more difficult to evaluate the peacetime European military balance, and their potential effects on the battlefield are uncertain. Overall, the main U.S. priorities are to reduce the vulnerability of theater-based nuclear retaliatory forces and to repair the political damage caused by friction over economic and security policies. The United States must constantly seek a middle ground between policies which indicates to its allies neither entrapment nor abandonment.

Nuclear proliferation is most problematic in the Middle East, where several threshold states are adversaries of Israel, a U.S. ally. It is in the Middle East that nuclear nonproliferation efforts need to be most concentrated. Germany and Japan are also problematic. If these countries acquire nuclear weapons, the United States needs to become their nuclear ally, as it did with France, rather than sever political ties.

Consider the concept of threat control in more detail. The United States and the Soviet Union have written, though not necessarily ratified, thirteen arms control agreements since 1959. These include agreements that preclude the deployment of nuclear weapons in particular regions, agreements on test constraints, agreements on deployment limitations, and confidence-building measures.[1] These categorizations can encourage us to think more broadly about arms control rather than restrict our definition of the concept to negotiated agreements that specify arms reductions.

Threat control would include reducing threats to retaliatory forces, which could be achieved in several ways. Limitations on the number of full-range flight tests for ICBMs that both the United States and the Soviet Union could conduct would lower confidence in the accuracy and reliability of these weapons. This lowering of confidence would in turn reduce the likelihood that decisionmakers would consider as a credible option a counterforce first strike using these weapons. Preannounced

1. The thirteen agreements are the Antarctic Treaty, the Outer Space Treaty, the Latin America Nuclear Free Zone Treaty, the Seabed Treaty, the Limited Test Ban Treaty, the Threshold Test Ban Treaty, the Agreement on Peaceful Nuclear Explosions, the Nuclear Nonproliferation Treaty, the SALT I agreements, the SALT II Treaty, the Hotline Agreement, the Accidental Measures Agreement, and the Prevention of Nuclear War Agreement. See *Arms Control and Disarmament Agreements: Texts and History of Negotiations* (Washington, D.C.: United States Arms Control and Disarmament Agency, 1980).

tests of identified systems that were restricted to specific test ranges chosen for ease of test monitoring would help reduce the risk of the Soviet Union breaking the agreement on short notice and conducting a rapid sequence of confidence tests. Moreover, simultaneous tests of more than one missile system could be banned to assist in test monitoring, and depressed-trajectory SLBM tests could be prohibited.[2] Reduction in threats to sea-based forces could be achieved by limiting the total number of deployed hunter-killer submarines and by prohibiting active trailing of submarines carrying ballistic missiles.

Weapon system modernization could be curtailed through limiting flight tests, restricting the number of new missiles and aircraft that could be deployed during a specified period, and restricting the upgrading of systems already deployed.

Crisis-avoidance measures might include not allowing submarines carrying ballistic missiles to enter specified patrol areas.

Threats to command, control, communication, and intelligence systems could be reduced by barring the testing or deployment of advanced weaponry that could destroy satellites and by agreeing not to interfere with national technical means of verification. (Despite general provisions to this effect in the SALT I agreements, this issue has continued to be a source of friction between the United States and the Soviet Union.)

Any unilateral step taken to reduce the vulnerability of a retaliatory system that simultaneously reduces the fear of preemption by the potential adversary is an act of threat control. Threat control need not be bilateral.

Nor need threat control be symmetrical, although asymmetry in this area would arouse intense domestic political controversy. An example of asymmetrical threat control would be restrictions on U.S. antisubmarine warfare behavior as compensation for reductions in Soviet heavy ICBMs. Identifying and removing one threat at a time is far more sensible and feasible than trying to redesign the forces of both sides.

A final way to control threat is through confidence-building measures such as establishing crisis management centers in Washington and Moscow. But these, it must be understood, would facilitate cooperative behavior only if both sides have the will to cooperate. They cannot in themselves alter the basic character of the relationship. The most

2. Depressed-trajectory SLBMs follow flight paths with lower apogees than missiles flying ballistic trajectories. They are designed to avoid early warning radars and could be used to attack bomber bases, missile silos, or submarines in port.

effective confidence-building measures would indeed be acts of mutual threat control. It is therefore recommended that public interest groups lend their support to specific threat reduction proposals rather than focus on comprehensive formulas that have little hope of being implemented and would have a negligible impact on the stability of the strategic balance of forces if they were.

The United States is locked into a protracted, indefinite global competition with the Soviet Union in which nuclear weapons play a key role. Resolution of this competition will be slow in coming, and progress in threat control will be difficult. Because of political and technological realities, we cannot escape the possibility of nuclear war. The best we can do is maintain secure retaliatory forces to minimize the risk of nuclear weapon use and establish whatever cooperative procedures we can with the Soviet Union so that nuclear war, once begun, could be ended as quickly as possible.

Despite the deeply competitive nature of U.S.-Soviet relations and the major strides in weapon system technology taken by both countries, fear of use of nuclear weapons seems to dominate the views of political leaders in both Washington and Moscow. The nuclear stalemate remains in place. It will take truly revolutionary technological innovation or a massive exercise of human stupidity before this stalemate is seriously threatened.

Index

STRATEGIC COMMAND AND CONTROL
Bruce G. Blair

During the past twenty-five years, U.S. strategists have argued that avoiding nuclear war depends on deterring a Soviet first strike by ensuring that U.S. forces could survive a surprise attack in numbers sufficient to inflict unacceptable damage in retaliation. U.S. military and political leaders have thus emphasized acquiring more powerful and accurate weaponry and providing better protection for it, while defense analysts have focused on assessing the relative strength and survivability of U.S. and Soviet forces. In the process neither has given sufficient attention to the vulnerability of the U.S. command, control, and communications system that would coordinate warning of an attack in progress and the response to it. In this study Bruce Blair examines accepted assumptions about mutual deterrence, force strength, and survivability, and concludes that the vulnerability of command, control, and communications not only precludes an effective retaliatory strike but also invites a preemptive Soviet first strike.

After summarizing the assumptions and evaluative methodology behind mainstream strategic theory, the study describes the current decentralized command and control system that, under conditions of surprise attack, could be unable to communicate with decisionmakers or with units responsible for executing the decisions. Blair traces in detail the development of the system over three decades; the attempts to improve it through the use of procedural guidelines, alternative and redundant communications channels, and survival tactics; and the continuing vulnerabilities from improved Soviet weapons and the environmental forces engendered by massive nuclear detonations. Blair also analyzes the probable effects of proposals by the Reagan administration to strengthen command, control, and communications systems and provides recommendations for further strengthening and for altering related policies, deployments, and strategies to improve the stability of deterrence.

Bruce G. Blair was a research assistant in the Brookings Foreign Policy Studies program; he is currently at the Department of Defense.

1985/c. 430 pp./cloth and paper

Jacket designed by Stephen Kraft